INHALT

Hauptaufsätze

Wir bitten, **Manuskriptsendungen** zweifach (Original und eine Kopie, sprachlich einwandfrei, Formeln mit Maschine oder in Druckschrift geschrieben) an folgende Anschrift zu richten:

Zeitschrift für Angewandte Mathematik und Mechanik
Institut für Mechanik der Akademie der Wissenschaften der DDR
DDR-1199 Berlin, Rudower Chaussee 5

Zu den Arbeiten, die als Hauptaufsätze bestimmt sind, ist auf gesondertem Blatt eine Zusammenfassung von 5 bis 10 Zeilen in englischer und (möglichst) deutscher und russischer Sprache beizufügen. Ausführliche Hinweise für die Autoren, um deren strikte Berücksichtigung gebeten wird, finden sich im Anschluß an das Inhaltsverzeichnis des Jahrganges **60** (1979).

Die Autoren erhalten von den Hauptaufsätzen 75, von den Kleinen Mitteilungen 25 Sonderdrucke ohne Berechnung, darüber hinaus weitere Sonderdrucke gegen Berechnung.

Der Verlag behält sich für alle Beiträge das Recht der Vervielfältigung, Vorbereitung und Übersetzung vor.

Bestellungen sind zu richten

- **in der DDR** an den Postzeitungsvertrieb unter Angabe der Kundennummer des Bestellenden oder an den AKADEMIE-VERLAG, DDR-1080 Berlin, Leipziger Str. 3–4
- **im sozialistischen Ausland** an eine Buchhandlung für fremdsprachige Literatur oder an den zuständigen Postzeitungsvertrieb
- **in der BRD und Berlin(West)** an eine Buchhandlung oder an die Auslieferungsstelle KUNST UND WISSEN, Erich Bieber OHG, D-7000 Stuttgart 1, Wilhelmstraße 4–6
- **in den übrigen westeuropäischen Ländern** an eine Buchhandlung oder an die Auslieferungsstelle KUNST UND WISSEN, Erich Bieber GmbH, CH-8008 Zürich/Schweiz, Dufourstraße 51
- **im übrigen Ausland** an den Internationalen Buch- und Zeitschriftenhandel; den Buchexport, Volkseigener Außenhandelsbetrieb der Deutschen Demokratischen Republik, DDR-7010 Leipzig, Postfach 160; oder an den AKADEMIE-VERLAG, DDR-1080 Berlin, Leipziger Straße 3–4

ZEITSCHRIFT FÜR ANGEWANDTE MATHEMATIK UND MECHANIK
Herausgeber und Chefredakteur: Prof. Dr. Günter Schmidt. Redaktion: Dr. Winfried Heinrich, Dr. Horst Weinert, Dipl.-Math. Friedhild Dudel, Helga Rühl, Institut für Mechanik der Akademie der Wissenschaften der DDR. Verlag: Akademie-Verlag, DDR-1080 Berlin, Leipziger Straße 3–4; Fernruf: 2236221 oder 2236229; Telex-Nr.: 114420; Bank: Staatsbank der DDR, Berlin, Kto.-Nr. 6836-26-20712. Anschrift der Redaktion: Institut für Mechanik der Akademie der Wissenschaften, DDR-1199 Berlin, Rudower Chaussee 5; Fernruf: 6802841

Veröffentlicht unter der Lizenznummer 1282 des Presseamtes beim Vorsitzenden des Ministerrates der Deutschen Demokratischen Republik. Gesamtherstellung: VEB Druckerei „Thomas Müntzer", 5820 Bad Langensalza. Erscheinungsweise: Die Zeitschrift für Angewandte Mathematik und Mechanik erscheint monatlich. Die 12 Hefte eines Jahres einschließlich Tagungsheft bilden einen Band. Bezugspreis je Band 360,— M zuzüglich Versandspesen (Preis für die DDR 216,— M). Bezugspreis je Heft 30,— M (Preis für die DDR 18,— M). Bestellnummer dieses Heftes: 1009/61/7.

AN (EDV) 35937

ZAMM 61, 277 – 292 (1981)

A. D. DE PATER

The General Theory of the Motion of a Single Wheelset Moving through a Curve with Constant Radius and Cant

Die Arbeit behandelt die Bewegung eines Radsatzes auf einer Kurve mit konstantem Radius und konstanter Überhöhung. Die Untersuchung schließt sowohl die stationäre als auch die parasitäre Bewegung ein. Die allgemeinen Bewegungsgleichungen werden sowohl mittels der Lagrangeschen Methode als auch durch Formulierung des Schwerpunkt- und Drallsatzes gewonnen. Wenn der Kurvenradius gegen unendlich strebt, können die Bewegungsgleichungen als exakt betrachtet werden. Die Gleichungen sind sehr kompliziert und sollten, bevor man ihre Lösungen untersucht, vereinfacht werden, indem man berücksichtigt, daß die Verschiebungen des Radsatzes und der Kontaktpunkte stets klein sind gegenüber der Gleisspurweite und daß die Winkelverdrehungen klein gegenüber eins sind. Die Abfassung einer solchen Theorie erster Ordnung wird der Gegenstand einer weiteren Publikation sein.

Paper deals with the problem of the motion of a railway wheelset through a curve with constant radius and cant. The investigation includes both the stationary and the parasitic movement. General equations of motion are derived; both as Lagrange's equations and as the equations expressing the rate of change of the momentum and of the moment of momentum of the wheelset. For values of the curve radius which tend to infinity, the equations of motion can be considered as exact ones. The equations are very complicated and before solving them they should be reduced by taking into account that the wheelset and contact point displacements always remain small as compared with the track gauge, whereas the rotation angles (of the parasitic motion) of the wheelset remain small with respect to unity. The draft of such a first-order theory will be the subject of a forthcoming publication.

В работе рассматривается движение колёсного ската на кривой постоянного радиуса и постоянного наклона. Исследование включает как основное так и паразитическое движение. Общие уравнения движения получаются посредством метода Лагранжа, а также путём метода Ньютона-Эйлера. Если радиус кривой стремится к бесконечности, уравнения движения могут рассматриваться как точные. Уравнения очень сложны; прежде чем исследовать их решения, необходимо упрощать уравнения учётом того, что перемещения колёсного ската и точек контакта малы по сравнению с расстоянием между рельсами и что вращения малы по сравнению с единицей. Составление такой теории первого порядка будет темой дальнейшей публикации.

List of symbols

The numbers denote the sections where the symbols are used for the first time.

Symbol	Meaning	Section
a_{xyij}, a_{xzij}, a_{yzij}	functions of coordinates x_{ij}^*, y_{ij}^*, z_{ij}^*	3
b_{xyij}, b_{xzij}, b_{yzij}	do.	3
b	half-gauge of track	3
$\boldsymbol{b}_i^*$	moment of momentum of wheelset	9
f, f^*	rail and tyre profile functions	3
g	acceleration of gravity	6
i	index of wheelset	2
j	rail index (1 for right-hand, 2 for left-hand)	3
m_i	wheelset mass	6
m_{ij}	coefficient of matrix $\boldsymbol{M}$	9
n_{ij}, n_{ij}^*	directions of normal in contact point	3
o_i	origin of moving coordinate system	2
o_i^*	mass centre of wheelset	2
$\boldsymbol{p}_i$	momentum of wheelset	9
$\boldsymbol{q}_i$	displacement vector	2
r	wheel radius	3
r_0	curve radius	1
$\boldsymbol{r}_0$	vector in coordinate system (o_i^*, x_0, y_0, z_0)	2
r_i	vector in coordinate system (o_i, x_i, y_i, z_i)	2
$\boldsymbol{r}_i^*$	vector in coordinate system $(o_i^*, x_i^*, y_i^*, z_i^*)$	2
r_i^*	radius of a contact point	3
r_{ij}^*	radius of contact point A_{ij}	3
$\boldsymbol{r}_{ij}^*$	vector of contact point in coordinate system $(o_i^*, x_i^*, y_i^*, z_i^*)$	9
s_i	displacement of wheelset along curve	2
t_{xij}, t_{yij}	tangential directions in contact point	3
t_{xij}^*, t_{yij}^*	do.	3
v_i, w_i	displacements of wheelset	2
x_0, y_0, z_0	axes of coordinate system with origin o_i^*	2
x_i, y_i, z_i	do. of inclined system with origin o_i	2
x_0^*, y_0^*, z_0^*	do. do. with origin o_i^*	2
x_i^*, y_i^*, z_i^*	do. do. do.	2
x_{ij}, y_{ij}, z_{ij}	coordinates of contact point	3
x_{ij}^*, y_{ij}^*, z_{ij}^*	do.	3
x_{ij}^{**}, y_{ij}^{**}, z_{ij}^{**}	do. with respect to system rotated about angle θ_{ij}	3
A_{ij}	contact point	3
$\boldsymbol{F}_i$	resultant of external forces applied at wheelset in o_i^*	6
F_{xi}, F_{yi}, F_{zi}	components of $\boldsymbol{F}_i$	6
$\boldsymbol{G}_{0i}$	matrix of direction cosines between $(\omega, \xi, \eta, \zeta)$ and (o_i, x_i, y_i, z_i)	2
$\boldsymbol{G}_i$	matrix of direction cosines between (o_i^*, x_0, y_0, z_0) and $(o_i^*, x_i^*, y_i^*, z_i^*)$	2
$\boldsymbol{G}_{ij}$, $\boldsymbol{G}_{ij}^*$	do. for normal and tangential directions	3

J_i, J_{yi}	wheelset inertia moments	7
$\boldsymbol{M}$	matrix	9
$\boldsymbol{M}_i$	resulting moment of external forces applied at wheelset	6
M_{xi}, M_{yi}, M_{zi}	components of $\boldsymbol{M}_i$	6
N_{ij}	normal force in contact point	6
O	curve centre	2
O_{ij}	origin of auxiliary system	3
P	power	8
P_{ex}	power of external forces applied at wheelset	8
P_{in}	power of contact point forces	8
Q_i, Q_{si} etc.	work coefficients	6
T	kinetic energy	7
T_{xij}, T_{yij}	tangential forces	6
U	potential energy	6
$\boldsymbol{V}_i$	translational velocity of o_i^*	5
$\mathrm{W}_{nij}, \mathrm{W}_{txij}, \mathrm{W}_{tyij}$	velocities in contact point	5
X_{ij}, Y_{ij}, Z_{ij}	coordinates in auxiliary system	3
X_{ij}, Y_{ij}, Z_{ij}	contact point forces	9
$X_{ij}^*, Y_{ij}^*, Z_{ij}^*$	do.	9
β_{ij}	resultant conicity angle	3
γ_0	conicity in central position	3
γ_i, γ_i^*	general conicity angles	3
$\gamma_{ij}, \gamma_{ij}^*$	conicity angles	3
δ	variation	4
$\delta\boldsymbol{r}_{i0}$	virtual displacement of mass centre o_i^*	4
$\delta\boldsymbol{s}_{ij}^*$	virtual displacement in contact point	4
$\delta s_{nij}, \delta s_{txij}, \delta s_{tyij}$	components of do.	4
δW	virtual work	6
δW_{ex}	virtual work of external forces applied at wheelset	6
δW_{in}	virtual work of contact point forces	6
$\delta\boldsymbol{\theta}_i$	virtual rotation of wheelset with respect to system (o_i, x_i, y_i, z_i)	4
$\delta\boldsymbol{\theta}_i^*$	virtual rotation of wheelset with respect to system $(o_i^*, x_i^*, y_i^*, z_i^*)$	4
$\delta\boldsymbol{\theta}^{*p}$	virtual rotation of parasitic movement	4
$\delta\boldsymbol{\theta}_{i0}^*$	virtual rotation of system $(o_i^*, x_i^*, y_i^*, z_i^*)$	4
$\delta\boldsymbol{\theta}_{0i}^J$	virtual rotation of fundamental movement	4
$\delta\bar{\boldsymbol{\theta}}_i^*$	matrix of virtual rotation	4
$\eta. \zeta$	coordinates in fixed system	2
η_{0i}, ζ_{0i}	coordinates of wheelset mass centre	2
η_i, ζ_i	coordinates in moving system	2
θ_i	wheelset rotation about axis of revolution	2
θ_{ij}	angle determining contact point	3
$\dot{\boldsymbol{\theta}}_i^*$	angular velocity of wheelset	4
$\dot{\boldsymbol{\theta}}_{i0}^*$	angular velocity of system $(o_i^*, x_i^*, y_i^*, z_i^*)$	4
ξ	coordinate in fixed system	2
ξ_{0i}	coordinate of wheelset mass centre	2
ξ_i	coordinate in moving system	2
$\boldsymbol{\varrho}_{0i}$	radius vector of origin o_i	2
$\boldsymbol{\varrho}_i$	vector of point of wheelset	2
$\boldsymbol{\varrho}_{i0}$	vector of mass centre	4
v_{xij}, v_{yij}	creepage components	10
φ_0	cant angle	2
φ_i, ψ_i	wheelset rotations	2
φ_{ij}	spin	10
ψ_{0i}	rotatory angle of system $(o_i, \xi_i, \eta_i, \zeta_i)$	2
ψ_{ij}	angle of auxiliary system $(O_i, \Xi_{ij}, H_{ij}, \mathfrak{Z}_{ij})$	3
ω	origin of fixed coordinate system	2
$\boldsymbol{\omega}_i$	wheelset angular velocity with respect to system (o_i, x_i, y_i, z_i)	5
$\boldsymbol{\omega}_i^*$	wheelset angular velocity with respect to system $(o_i^*, x_i^*, y_i^*, z_i^*)$	5
$\boldsymbol{\omega}_{i0}$	angular velocity of system (o_i, x_i, y_i, z_i)	5
$\boldsymbol{\omega}_{i0}^*$	angular velocity of system $(o_i^*, x_i^*, y_i^*, z_i^*)$	5
$\omega_{nij}. \omega_{txij}, \omega_{tyij}$	components of wheel angular velocity in contact point	5
$\Xi_{ij}, H_{ij}, \mathfrak{Z}_{ij}$	coordinates in auxiliary system	3

1. Introduction

Railway technology dates back to more than two centuries; nevertheless, until recently the dynamics of the interaction between track and vehicle remained a relatively obscure phenomenon, which during last years gives rise to an ever swelling stream of publications.

The problem of the motion of a railway through a curve has been investigated continuously during the last century in Germany and Switzerland, countries where a large part of the railway track is laid in curves. Most of the relevant publications deal with the stationary motion along a curved track with constant radius but also attention was paid to the running into curves and out of curves. HEUMANN [1] published an interesting review paper, whereas BORGEAUD [2] and MÜLLER [3] investigated more fundamental aspects of the problem; BORGEAUD confined himself partially and MÜLLER completely to the motion of the single wheelset.

Since about 1920 English and French scientists investigated the parasitic motion of wheelsets and complete vehicles on a tangent track.

The combination of the two aforesaid problems, i.e. the parasitic motion of single wheelsets and complete vehicles when running through a curve, has not yet been investigated much in detail. The present publication is devoted to such an investigation. We shall confine ourselves to a single wheelset and a curve with constant radius and cant; we shall show that a very general theory for the motion can be set up. This theory is an exact theory, irrespective of one simplification: we omitted second and higher degree terms in r_0^{-1} (r_0 being the curve radius), which reduces the equations considerably and, on the other hand, has nearly no influence on their accuracy, even in the case of streetcar systems with large curvatures.

The idea of setting up such equations of motion as are shown in the present report, was originated by ALDINGTON [4]. Author has applied it previously to the problem of the single wheelset running on a tangent track [5, 6]; for this case we can write down equations of motion which are completely exact.

For a practical investigation it is advantageous to simplify the equations of motion before trying to solve them. This can be done in a way which was used already previously [4, 5] for the motion on a tangent track, for which both a so-called first-order theory and a completely linear theory were drawn up. For the curve problem the analogous part of the investigation will be the subject of a forthcoming publication. Another problem which will have to be investigated in future, is the problem of the motion on a (tangent or curved) track with rails which present small irregularities.

The forces in the point where a rail and a wheel contact each other, depend on the so-called creepage and spin quantities. The relationship has been investigated at full length by KALKER [7].

The sections into which the present paper has been divided, agree largely with the sections of chapter 2 of the publications [5] and [6] on the motion of a wheelset on a tangent track. For the sake of clarity the paper has been made complete in itself. First the geometrical and kinematical properties of the contact between wheel and rail are investigated. Secondly, the forces are scrutinized and the expressions for their work coefficients are derived; moreover, the kinetic energy of the system is calculated. At last the equations of motion are written down as LAGRANGE's equations of the system; they are checked by deriving them another time, viz. by means of NEWTON's laws.

2. Geometrical description of the motion of the wheelset

The fundamental motion of the wheelset is described by means of the path of its centre of gravity. This is a circle (fig. 1) with radius r_0, situated in a horizontal plane, with centre O. The distance ωo_i along the circle is indicated by s_i, o_i being the origin of the coordinate system $(o_i, \xi_i, \eta_i, \zeta_i)$, $o_i\xi_i$ being the tangent to the circle and $o_i\zeta_i$ being vertical; the plane $\xi_i = 0$ always contains the mass centre o_i^* of the wheelset. All the coordinate systems in this paper are assumed to be right-handed. We also introduce the fixed coordinate system $(\omega, \xi, \eta, \zeta)$ and the inclined coordinate system (o_i, x_i, y_i, z_i), where $o_ix_i \equiv o_i\xi_i$ and φ_0 is the cant angle: the angle between $o_i\eta_i$ and o_iy_i. The plane $x_i = 0$ intersects both of the rails according to a profile curve and the two profile curves are symmetrically situated with respect to o_iz_i.

The angle ψ_{0i} between $\omega\xi$ and $o_i\xi_i$ is equal to

$$\psi_{0i} = s_i/r_0 \,. \tag{2-1}$$

Thus the coordinates of the origin o_i are

$$\xi_{0i} = r_0 \sin\frac{s_i}{r_0}, \qquad \eta_{0i} = r_0\left(1 - \cos\frac{s_i}{r_0}\right), \qquad \zeta_{0i} = 0\,. \tag{2-2}$$

For the relation between the coordinates of the systems $(\omega, \xi, \eta, \zeta)$, $(o_i, \xi_i, \eta_i, \zeta_i)$ and (o_i, x_i, y_i, z_i) we easily find

$$\boldsymbol{\varrho}_i = \boldsymbol{\varrho}_{0i} + \boldsymbol{G}_{0i}\boldsymbol{r}_i\,, \tag{2-3}$$

where

$$\tilde{\boldsymbol{\varrho}}_i = (\xi_i, \eta_i, \zeta_i)\,, \qquad \tilde{\boldsymbol{\varrho}}_{0i} = (\xi_{0i}, \eta_{0i}, 0)\,, \qquad \tilde{\boldsymbol{r}}_i = (x_i, y_i, z_i) \tag{2-4}$$

and

$$\boldsymbol{G}_{0i} = \begin{pmatrix} \cos\psi_{0i} & -\sin\psi_{0i} & 0 \\ \sin\psi_{0i} & \cos\psi_{0i} & 0 \\ 0 & 0 & 1 \end{pmatrix} \begin{pmatrix} 1 & 0 & 0 \\ 0 & \cos\varphi_0 & -\sin\varphi_0 \\ 0 & \sin\varphi_0 & \cos\varphi_0 \end{pmatrix}, \tag{2-5}$$

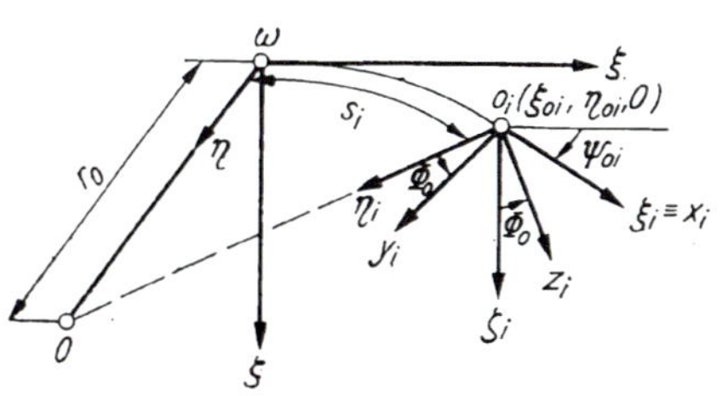

Fig. 1. The fundamental motion of the wheelset (Read φ and ψ instead of Φ and Ψ)

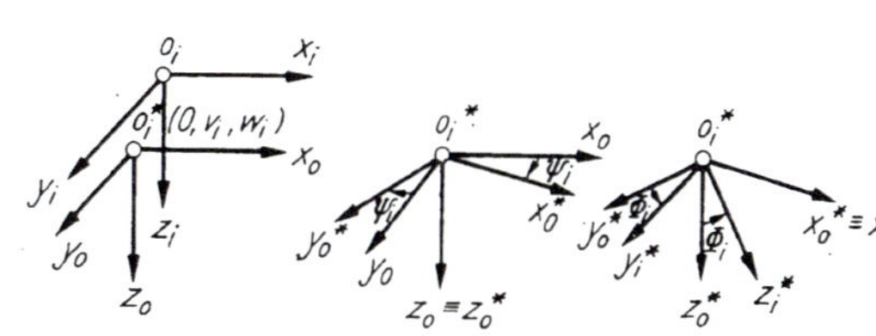

Fig. 2. The parasitic motion of the wheelset (Read φ and ψ instead of Φ and Ψ)

viz.

$$\boldsymbol{G}_{0i} = \begin{pmatrix} \cos\psi_{0i} & -\cos\varphi_0 \sin\psi_{0i} & \sin\varphi_0 \sin\psi_{0i} \\ \sin\psi_{0i} & \cos\varphi_0 \cos\psi_{0i} & -\sin\varphi_0 \cos\psi_{0i} \\ 0 & \sin\varphi_0 & \cos\varphi_0 \end{pmatrix}. \tag{2-6}$$

In the fundamental motion the centre of gravity o_i^* of the wheelset coincides with o_i and its axis of revolution with $o_i y_i$. To the fundamental motion also belongs the rotation of the wheelset when it is in its central position; this rotation will be discussed in combination with the parasitic motion.

The latter motion is more complicated. It is illustrated by fig. 2; it is described by the displacements v_i, w_i and the angles φ_i, θ_i, ψ_i. The quantities v_i, w_i, φ_i, ψ_i are indicated in fig. 2; they define the coordinate system $(o_i^*, x_i^*, y_i^*, z_i^*)$, the axis $o_i^* y_i^*$ coinciding with the axis of revolution of the wheelset. The angle θ_i defines the rotation of the wheelset about its axis of revolution. According to fig. 2 we can write:

$$\tilde{\boldsymbol{q}} = (o, v_i, w_i)\,, \qquad \tilde{\boldsymbol{r}}_0 = (x_0, y_0, z_0) \tag{2-7}$$

and

$$\boldsymbol{r}_i = \boldsymbol{q}_i + \boldsymbol{r}_0; \tag{2-8}$$

moreover:

$$\boldsymbol{r}_0 = \boldsymbol{G}_i \boldsymbol{r}_i^* \tag{2-9}$$

with

$$\tilde{\boldsymbol{r}}_i^* = (x_i^*, y_i^*, z_i^*)\,, \qquad \boldsymbol{G}_i = \begin{pmatrix} \cos\psi_i & -\cos\varphi_i \sin\psi_i & \sin\varphi_i \sin\psi_i \\ \sin\psi_i & \cos\varphi_i \cos\psi_i & -\sin\varphi_i \cos\psi_i \\ 0 & \sin\varphi_i & \cos\varphi_i \end{pmatrix}. \tag{2-10}$$

Note the conformity between $\boldsymbol{G}_{0i}$ (6) and $\boldsymbol{G}_i$ (10). We also shall need the expression

$$\boldsymbol{G}_i^{-1} = \tilde{\boldsymbol{G}}_i = \begin{pmatrix} \cos\psi_i & \sin\psi_i & 0 \\ -\cos\varphi_i \sin\psi_i & \cos\varphi_i \cos\psi_i & \sin\varphi_i \\ \sin\varphi_i \sin\psi_i & -\sin\varphi_i \cos\psi_i & \cos\varphi_i \end{pmatrix}. \tag{2-11}$$

Combining (3), (8) and (9) yields

$$\tilde{\boldsymbol{G}}_{0i}\boldsymbol{Q}_i = \tilde{\boldsymbol{G}}_{0i}\boldsymbol{Q}_{0i} + \boldsymbol{q}_i + \boldsymbol{G}_i \boldsymbol{r}_i^*\,. \tag{2-12}$$

3. The constraints between track and wheelset

Usually a wheel contacts the rail along which it moves, in one point. Often the rail and wheel profiles are such that two contact points may occur; we shall leave such a double contact out of consideration. Then there exist two constraint relations between the quantities v_i, w_i, φ_i and ψ_i, which we now shall derive.

We introduce the coordinates x_{ij}, y_{ij}, z_{ij} of the contact points with respect to the system (o_i, x_i, y_i, z_i), the corresponding coordinates x_{ij}^*, y_{ij}^*, z_{ij}^* with respect to the system $(o_i^*, x_i^*, y_i^*, z_i^*)$, the profile functions f and f^* and the angles γ_{ij} and γ_{ij}^* of the profiles in the contact points: the angles which define the conicities.

The surfaces of the rails can be described by means of the coordinate systems $(o_i, \xi_i, \eta_i, \zeta_i)$ and (o_i, x_i, y_i, z_i) of fig. 1 and the systems $(O_{ij}, \Xi_{ij}, H_{ij}, \mathfrak{Z}_{ij})$ and $(O_{ij}, X_{ij}, Y_{ij}, Z_{ij})$, shown in fig. 3. This figure refers to the right-hand rail surface ($j = 1$); the left-hand surface ($j = 2$) can be tackled in an analogous way.

The plane $\Xi_{i1} = 0$ is vertical and contains both the contact point A_{i1} and the curve centre O. The axes $O_{i1}\Xi_{i1}$ and $O_{i1}H_{i1}$ are horizontal. The angle between the axes $o_i\eta_i$ and $O_{i1}H_{i1}$ is equal to ψ_{ij}. The coordinate system $(O_{i1}, X_{i1}, Y_{i1}, Z_{i1})$ is introduced in such a way that $O_{i1}X_{i1} \equiv O_{i1}\Xi_{i1}$, whereas the angle between $O_{i1}H_{i1}$ and $O_{i1}Y_{i1}$ is equal to the cant angle φ_0.

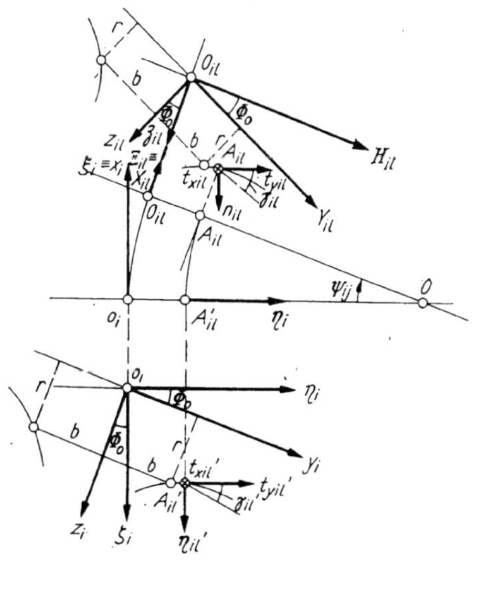

Fig. 3. The rail surfaces
(Read 1 and φ instead of l and Φ)

In the contact point A_{i1} we can indicate the normal direction n_{i1} (pointing inwards to the rail), the tangential direction t_{xi1}, parallel to the axis $O_{i1}X_{i1}$ (pointing into the positive direction of the axis) and the tangential direction t_{yi1}, perpendicular to n_{i1} and t_{xi1}; also t_{xij}, t_{yij} and n_{ij} form a right-handed system.

In the plane $x_i = 0$ we can draw its intersection curve with the rail surface: the rail profile curve. The point A'_{i1} is situated on this curve in such a way that $A_{i1}A'_{i1}$ is a parallel circle on the rail surface.

For the various coordinates we find the following relations:

$$\begin{pmatrix}\xi_i\\ \eta_i\\ \zeta_i\end{pmatrix} = \begin{pmatrix}1 & 0 & 0\\ 0 & \cos\varphi_0 & -\sin\varphi_0\\ 0 & \sin\varphi_0 & \cos\varphi_0\end{pmatrix}\begin{pmatrix}x_i\\ y_i\\ z_i\end{pmatrix}, \quad \begin{pmatrix}\Xi_{ij}\\ H_{ij}\\ \mathfrak{Z}_{ij}\end{pmatrix} = \begin{pmatrix}1 & 0 & 0\\ 0 & \cos\varphi_0 & -\sin\varphi_0\\ 0 & \sin\varphi_0 & \cos\varphi_0\end{pmatrix}\begin{pmatrix}X_{ij}\\ Y_{ij}\\ Z_{ij}\end{pmatrix} \tag{3-1}$$

and

$$\begin{pmatrix}\xi_i\\ \eta_i\\ \zeta_i\end{pmatrix} = \begin{pmatrix}r_0\sin\psi_{ij}\\ r_0(1-\cos\psi_{ij})\\ 0\end{pmatrix} + \begin{pmatrix}\cos\psi_{ij} & -\sin\psi_{ij} & 0\\ \sin\psi_{ij} & \cos\psi_{ij} & 0\\ 0 & 0 & 1\end{pmatrix}\begin{pmatrix}\Xi_{ij}\\ H_{ij}\\ \mathfrak{Z}_{ij}\end{pmatrix}. \tag{3-2}$$

Eliminating (ξ_i, η_i, ζ_i) and $(\Xi_{ij}, H_{ij}, \mathfrak{Z}_{ij})$ from (1) and (2) yields

$$\begin{aligned}\begin{pmatrix}x_i\\ y_i\\ z_i\end{pmatrix} &= \begin{pmatrix}1 & 0 & 0\\ 0 & \cos\varphi_0 & \sin\varphi_0\\ 0 & -\sin\varphi_0 & \cos\varphi_0\end{pmatrix}\begin{pmatrix}r_0\sin\psi_{ij}\\ r_0(1-\cos\psi_{ij})\\ 0\end{pmatrix} +\\ &+ \begin{pmatrix}\cos\psi_{ij} & -\cos\varphi_0\sin\psi_{ij} & \sin\varphi_0\sin\psi_{ij}\\ \cos\varphi_0\sin\psi_{ij} & \cos^2\varphi_0\cos\psi_{ij}+\sin^2\varphi_0 & \cos\varphi_0\sin\varphi_0(1-\cos\psi_{ij})\\ -\sin\varphi_0\sin\psi_{ij} & \cos\varphi_0\sin\varphi_0(1-\cos\psi_{ij}) & \cos^2\varphi_0+\sin^2\varphi_0\cos\psi_{ij}\end{pmatrix}\begin{pmatrix}X_{ij}\\ Y_{ij}\\ Z_{ij}\end{pmatrix}.\end{aligned} \tag{3-3}$$

For the contact point A_{ij} we have

$$x_i = x_{ij}, \quad y_i = y_{ij}, \quad z_i = z_{ij}, \quad X_{ij} = 0, \tag{3-4}$$

so that (3) reduces to

$$\left.\begin{aligned} x_{ij} &= r_0\sin\psi_{ij} - Y_{ij}\cos\varphi_0\sin\psi_{ij} + Z_{ij}\sin\varphi_0\sin\psi_{ij},\\ y_{ij} &= r_0(1-\cos\psi_{ij})\cos\varphi_0 + Y_{ij}(\cos^2\varphi_0\cos\psi_{ij}+\sin^2\varphi_0) + Z_{ij}\cos\varphi_0\sin\varphi_0(1-\cos\psi_{ij}),\\ z_{ij} &= -r_0(1-\cos\psi_{ij})\sin\varphi_0 + Y_{ij}(1-\cos\psi_{ij})\cos\varphi_0\sin\varphi_0 + Z_{ij}(\cos^2\varphi_0+\sin^2\varphi_0\cos\psi_{ij}).\end{aligned}\right\} \tag{3-5}$$

When ψ_{ij}, Y_{ij} and Z_{ij} are given, x_{ij}, y_{ij} and z_{ij} are determined by (5).

For the right-hand rail we can represent the profile curve by

$$z_i = f(y_i), \tag{3-6a}$$

so that the conicity angle γ_i is defined by

$$\operatorname{tg}\gamma_i = -\frac{df}{dy_i}. \tag{3-6b}$$

Thus we have for both contact points

$$Z_{ij} = f(\pm Y_{ij}), \quad \operatorname{tg}\gamma_{ij} = -\left(\frac{df}{dy_i}\right)_{y_i=\pm Y_{ij}}, \tag{3-7}$$

the upper sign being valid for $j = 1$ and the lower one for $j = 2$. For the central position we have

$$y_{ij} = \pm b, \quad z_{ij} = r, \quad \gamma_{ij} = \gamma_0 \tag{3-8}$$

and

$$r = f(b), \quad \operatorname{tg}\gamma_0 = -\left(\frac{df}{dy_i}\right)_{y_i=b}, \tag{3-9}$$

$2b$ being the track gauge and r the wheel radius in the central position.

The relations (5) become much more simple when $|\psi_{ij}| \ll 1$, so that we can put

$$\cos\psi_{ij} = 1, \quad \sin\psi_{ij} = \psi_{ij}. \tag{3-10}$$

In that case we have

$$x_{ij} = (r_0 - Y_{ij}\cos\varphi_0 + Z_{ij}\sin\varphi_0)\,\psi_{ij}, \quad y_{ij} = Y_{ij}, \quad z_{ij} = Z_{ij}; \tag{3-11}$$

moreover, then $|Y_{ij}| \ll r_0$, $|Z_{ij}| \ll r_0$, so that we may write

$$\psi_{ij} = x_{ij}/r_0.$$

Because of (7) we also have

$$z_{ij} = f(\pm y_{ij}), \quad \operatorname{tg}\gamma_{ij} = -\left(\frac{df}{dy_i}\right)_{y_i=\pm y_{ij}}. \tag{3-13}$$

We shall also need the direction cosines of the directions n_{ij}, t_{xij}, t_{yij}. We easily find the matrix of the direction cosines cos (n_{ij}, X_{ij}) etc.:

cos	X_{ij}	Y_{ij}	Z_{ij}
n_{ij}	0	$\pm \sin \gamma_{ij}$	$\cos \gamma_{ij}$
t_{xij}	1	0	0
t_{yij}	0	$\cos \gamma_{ij}$	$\mp \sin \gamma_{ij}$

(3-14)

Now (3) and (12) yield

cos	x_i	y_i	z_i
X_{ij}	1	$x_{ij} r_0^{-1} \cos \varphi_0$	$-x_{ij} r_0^{-1} \sin \varphi_0$
Y_{ij}	$-x_{ij} r_0^{-1} \cos \varphi_0$	1	0
Z_{ij}	$x_{ij} r_0^{-1} \sin \varphi_0$	0	1

; (3-15)

premultiplying this matrix with the previous one, we obtain

cos	x_i	y_i	z_i
n_{ij}	$\mp x_{ij} r_0^{-1} \sin \beta_{ij}$	$\pm \sin \gamma_{ij}$	$\cos \gamma_{ij}$
t_{xij}	1	$x_{ij} r_0^{-1} \cos \varphi_0$	$-x_{ij} r_0^{-1} \sin \varphi_0$
t_{yij}	$-x_{ij} r_0^{-1}$	$\cos \gamma_{ij}$	$\mp \sin \gamma_{ij}$

, (3-16)

where

$$\beta_{ij} = \gamma_{ij} \mp \varphi_0 \,. \tag{3-17}$$

The gist of the previous calculations is that we retained only terms with r_0^{-1} and omitted all terms with r_0^{-1} to the second and higher degree. Let alone this simplification the theory, enunciated in the present report, can be considered as an exact theory.

We now consider the wheelset. We designate the distance of the contact point A_{i1} to the axis $o_i^* y_i^*$ by r_i^*. Then we can write for the wheel profile curve in a way, analogous to (6a) and (6b):

$$r_i^* = f^*(y_i^*)\,, \qquad \operatorname{tg} \gamma_i^* = -\frac{\mathrm{d}f^*}{\mathrm{d}y_i^*}\,. \tag{3-18}$$

Thus for both contact points the relations

$$r_{ij}^* = f^*(\pm y_{ij}^*)\,, \qquad \operatorname{tg} \gamma_{ij}^* = -\left(\frac{\mathrm{d}f}{\mathrm{d}y_i^*}\right)_{y_i^* = \pm y_{ij}^*} \tag{3-19}$$

hold. For the central position we have

$$y_{ij}^* = \pm b\,, \qquad r_{ij}^* = r\,, \qquad \gamma_{ij}^* = \gamma_0 \tag{3-20}$$

and

$$r = f^*(b)\,, \qquad \operatorname{tg} \gamma_0 = -\left(\frac{\mathrm{d}f^*}{\mathrm{d}y_i^*}\right)_{y_i^* \, b}\,. \tag{3-21}$$

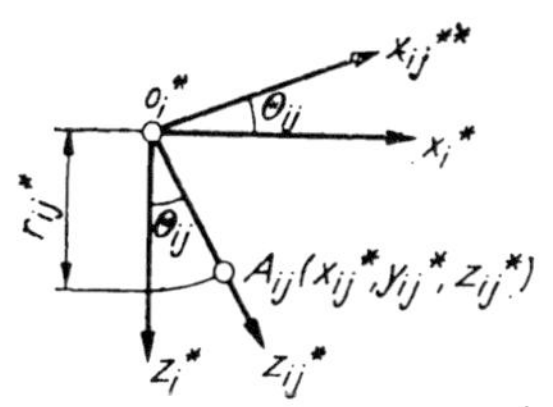

Fig. 4. The position of a contact point on a wheel

From fig. 4 we find the relation

$$x_{ij}^{*2} + z_{ij}^{*2} = r_{ij}^{*2}\,. \tag{3-22}$$

We introduce the coordinate system $(o_i^*, x_{ij}^{**}, y_{ij}^{**}, z_{ij}^{**})$, where $o_i^* y_i^* \equiv o_i^* y_{ij}^{**}$ and the contact point A_{ij} lies in the plane $x_{ij}^{**} = 0$. When we call θ_{ij} the angle between the axes $o_i^* x_i^*$ and $o_i^* x_{ij}^{**}$ (fig. 4), so that

$$\cos \theta_{ij} = z_{ij}^*/r_{ij}^*\,, \qquad \sin \theta_{ij} = x_{ij}^*/r_{ij}^*\,, \qquad \operatorname{tg} \theta_{ij} = x_{ij}^*/z_{ij}^*\,, \tag{3-23}$$

the relation with the system $(o_i^*, x_i^*, y_i^*, z_i^*)$ appears to read

$$\begin{pmatrix} x_i^* \\ y_i^* \\ z_i^* \end{pmatrix} = \begin{pmatrix} \cos \theta_{ij} & 0 & \sin \theta_{ij} \\ 0 & 1 & 0 \\ -\sin \theta_{ij} & 0 & \cos \theta_{ij} \end{pmatrix} \begin{pmatrix} x_{ij}^{**} \\ y_{ij}^{**} \\ z_{ij}^{**} \end{pmatrix}. \tag{3-24}$$

Then we are able to find the direction cosines of the normal n^*_{ij} in a contact point, which is taken positive when it points outwards with respect to the wheel. In agreement with (14) we find

cos	x^{**}_{ij}	y^{**}_{ij}	z^{**}_{ij}
n^*_{ij}	0	$\pm \sin\gamma^*_{ij}$	$\cos\gamma^*_{ij}$
t^*_{xij}	1	0	0
t^*_{yij}	0	$\cos\gamma^*_{ij}$	$\mp \sin\gamma^*_{ij}$

, (3-25)

so that, owing to (23) and (24), the matrix $\tilde{G}^*_{ij}$ is the matrix

cos	x^*_i	y^*_i	z^*_i
n^*_{ij}	$x^*_{ij} r^{*-1}_{ij} \cos\gamma^*_{ij}$	$\pm \sin\gamma^*_{ij}$	$z^*_{ij} r^{*-1}_{ij} \cos\gamma^*_{ij}$
t^*_{xij}	$z^*_{ij} r^{*-1}_{ij}$	0	$-x^*_{ij} r^{*-1}_{ij}$
t^*_{yij}	$\mp x^*_{ij} r^{*-1}_{ij} \sin\gamma^*_{ij}$	$\cos\gamma^*_{ij}$	$\mp z^*_{ij} r^{*-1}_{ij} \sin\gamma^*_{ij}$

. (3-26)

The constraint relations now can be found from two considerations: the coordinates of the contact points have to be in agreement with (2-12) and, moreover, the normal on a rail surface has to coincide with the normal on the contacting wheel surface.

Evaluating the first consideration we have to put into (2-12):

$$\tilde{G}_{0i}(\varrho_i - \varrho_{0i}) = \begin{pmatrix} x_{ij} \\ y_{ij} \\ z_{ij} \end{pmatrix}; \tag{3-27}$$

then we find, in connection with (2-7) and (2-10) and replacing x^*_i, y^*_i, z^*_i by $x^*_{ij}, y^*_{ij}, z^*_{ij}$:

$$\left.\begin{aligned} x_{ij} &= x^*_{ij}\cos\psi_i - a_{yzij}\sin\psi_i\,, \\ y_{ij} &= v_i + x^*_{ij}\sin\psi_i + a_{yzij}\cos\psi_i\,, \\ z_{ij} &= w_i + b_{yzij}\,. \end{aligned}\right\} \tag{3-28}$$

Here we have introduced the displacements a_{yzij} and b_{yzij} which, together with four analogous quantities which we shall need below, are defined by

$$a_{xyij} = x^*_{ij}\cos\varphi_i\cos\psi_i - y^*_{ij}\sin\psi_i\,,\qquad b_{xyij} = x^*_{ij}\cos\varphi_i\sin\psi_i + y^*_{ij}\cos\psi_i\,, \tag{3-29a}$$

$$a_{xzij} = x^*_{ij}\sin\varphi_i\cos\psi_i + z^*_{ij}\sin\psi_i\,,\qquad b_{xzij} = -x^*_{ij}\sin\varphi_i\sin\psi_i + z^*_{ij}\cos\psi_i\,, \tag{3-29b}$$

$$a_{yzij} = y^*_{ij}\cos\varphi_i - z^*_{ij}\sin\varphi_i\,,\qquad b_{yzij} = y^*_{ij}\sin\varphi_i + z^*_{ij}\cos\varphi_i\,. \tag{3-29c}$$

The second consideration gives rise to the relation

$$\tilde{G}_i \begin{pmatrix} \cos(n_{ij}, x_i) \\ \cos(n_{ij}, y_i) \\ \cos(n_{ij}, z_i) \end{pmatrix} = \begin{pmatrix} \cos(n^*_{ij}, x^*_i) \\ \cos(n^*_{ij}, y^*_i) \\ \cos(n^*_{ij}, z^*_i) \end{pmatrix}, \tag{3-30}$$

$\tilde{G}_i$ being determined by (2-11). By means of (16) and (26) we find

$$\begin{aligned} &x^*_{ij}\cos\gamma^*_{ij} \mp r^*_{ij}\sin\gamma_{ij}\sin\psi_i = \mp x_{ij} r^*_{ij} r_0^{-1}\sin\beta_{ij}\cos\psi_i\,, \\ &z^*_{ij}\cos\gamma^*_{ij} + r^*_{ij}(\pm\sin\gamma_{ij}\sin\varphi_i\cos\psi_i - \cos\gamma_{ij}\cos\varphi_i) = \pm x_{ij} r^*_{ij} r_0^{-1}\sin\beta_{ij}\sin\varphi_i\sin\psi_i\,, \\ &\sin\gamma^*_{ij} - \sin\gamma_{ij}\cos\varphi_i\cos\psi_i \mp \cos\gamma_{ij}\sin\varphi_i = x_{ij} r_0^{-1}\sin\beta_{ij}\cos\varphi_i\sin\psi_i\,. \end{aligned} \tag{3-31}$$

These relations are not independent: eliminating x^*_{ij} and z^*_{ij} from the first and the second set and making use of (22) we find the third set. Thus we can consider the relations (31) as an independant set and leave (22) out of consideration; we shall do so henceforward.

We now shall strike a balance about the number of unknowns and the number of equations available. For the

- six coordinates x_{ij}, y_{ij}, z_{ij},
- six do. $x^*_{ij}, y^*_{ij}, z^*_{ij}$,
- two radii r^*_{ij},
- four angles $\gamma_{ij}, \gamma^*_{ij}$,
- four quantities a_{yzij}, b_{yzij},

i.e. 24 unknowns, we dispose on 24 equations, viz.:

- four equations (13),
- four do. (19),
- six do. (28),
- four do. (29c),
- six do. (31).

For the calculations to be enunciated in the next section we shall also need the direction cosines of (n_{ij}, t_{xij}, t_{yij}) with respect to the coordinate system (o_i^*, x_i^*, y_i^*, z_i^*). Postmultiplying (16) with G_i (2-10) yields the matrix $\widetilde{G}_{ij}$ according to the following scheme:

cos	x_{ij}^*	y_{ij}^*	z_{ij}^*
n_{ij}	$\pm\sin\gamma_{ij}\sin\psi_i$	$\pm\sin\gamma_{ij}\cos\varphi_i\cos\psi_i+\cos\gamma_{ij}\sin\varphi_i$	$\mp\sin\gamma_{ij}\sin\varphi_i\cos\psi_i+\cos\gamma_{ij}\cos\varphi_i$
t_{xij}	$\cos\psi_i$	$-\cos\varphi_i\sin\psi_i$	$\sin\varphi_i\sin\psi_i$
t_{yij}	$\cos\gamma_{ij}\sin\psi_i$	$\cos\gamma_{ij}\cos\varphi_i\cos\psi_i\mp\sin\gamma_{ij}\sin\varphi_i$	$-\cos\gamma_{ij}\sin\varphi_i\cos\psi_i\mp\sin\gamma_{ij}\cos\varphi_i$

$+\dfrac{x_{ij}}{r_0}$

cos	x_{ij}^*	y_{ij}^*	z_{ij}^*
n_{ij}	$\mp\sin\beta_{ij}\cos\psi_i$	$\pm\sin\beta_{ij}\cos\varphi_i\sin\psi_i$	$\mp\sin\beta_{ij}\sin\varphi_i\sin\psi_i$
t_{xij}	$\cos\varphi_0\sin\psi_i$	$\cos\varphi_0\cos\varphi_i\cos\psi_i-\sin\varphi_0\sin\varphi_i$	$-\cos\varphi_0\sin\varphi_i\cos\psi_i-\sin\varphi_0\cos\varphi_i$
t_{yij}	$-\cos\beta_{ij}\cos\psi_i$	$\cos\beta_{ij}\cos\varphi_i\sin\psi_i$	$-\cos\beta_{ij}\sin\varphi_i\sin\psi_i$

(3-32)

4. The virtual displacements

We shall now calculate the translational and angular displacements of the wheelset, caused by varying the coordinates s_i, v_i, w_i, φ_i, θ_i and ψ_i. From them the translational and angular velocities of the wheelset can be found and also its virtual displacements in the contact points; the latter quantities are needed for determining the work coefficients of the forces in the contact points.

For the mass centre of the wheelset the vector r_i^* is zero. Thus, according to (2-12), we can write, after introducing the vector ϱ_{i0} of the mass centre o_i^* with respect to the system (ω, ξ, η, ζ),

$$\varrho_{i0}=\varrho_{0i}+G_{0i}q_i \tag{4-1}$$

and

$$\delta\varrho_{i0}=\delta\varrho_{0i}+\delta G_{0i}q_i+G_{0i}\,\delta q_i\,,$$

viz.

$$\widetilde{G}_{0i}\,\delta\varrho_{i0}=\widetilde{\widetilde{G}}_{0i}\,\delta\overline{\varrho}_{0i}+\widetilde{G}_{0i}\,\delta G_{0i}q_i+\delta q_i\,. \tag{4-2}$$

Because of (2-5) we have

$$\widetilde{G}_{0i}\,\delta G_{0i}=\begin{pmatrix}0 & -\cos\varphi_0 & \sin\varphi_0\\ \cos\varphi_0 & 0 & 0\\ -\sin\varphi_0 & 0 & 0\end{pmatrix}\delta\psi_{0i} \tag{4-3}$$

and with (2-1), (2-2), (2-4) and (2-5) we obtain, putting

$$\delta r_{i0}=\begin{pmatrix}\delta x_{i0}\\ \delta y_{i0}\\ \delta z_{i0}\end{pmatrix}=\widetilde{G}_{0i}\,\delta\varrho_{i0}: \tag{4-4}$$

$$\left.\begin{aligned}\delta x_{i0}&=\delta s_i+r_0^{-1}(-v_i\cos\varphi_0+w_i\sin\varphi_0)\,\delta s_i\,,\\ \delta y_{i0}&=\delta v_i\,,\\ \delta z_{i0}&=\delta w_i\,.\end{aligned}\right\} \tag{4-5}$$

The virtual rotations of the fundamental movement without the rotation of the wheelset about its axis of revolution are found by means of fig. 5:

$$\delta\theta_{0xi}^f=0\,,\qquad \delta\theta_{0yi}^f=r_0^{-1}\sin\varphi_0\cdot\delta s_i\,,\qquad \delta\theta_{0zi}^f=r_0^{-1}\cos\varphi_0\cdot\delta s_i\,. \tag{4-6}$$

Those of the parasitic motion, including this rotation of the wheelset (to be indicated by $\delta\theta_i$) can be derived from fig. 6:

$$\delta\theta_{xi}^{p*}=\delta\varphi_i\,,\qquad \delta\theta_{yi}^{p*}=-\delta\theta_i+\sin\varphi_i\cdot\delta\psi_i\,,\qquad \delta\theta_{zi}^{p*}=\cos\varphi_i\cdot\delta\psi_i\,. \tag{4-7}$$

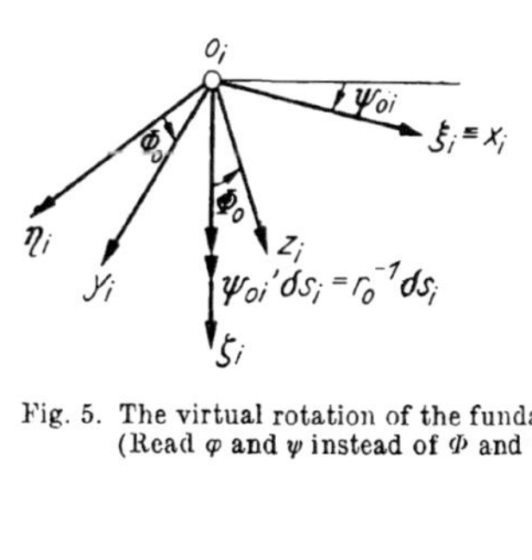

Fig. 5. The virtual rotation of the fundamental motion (Read φ and ψ instead of Φ and Ψ)

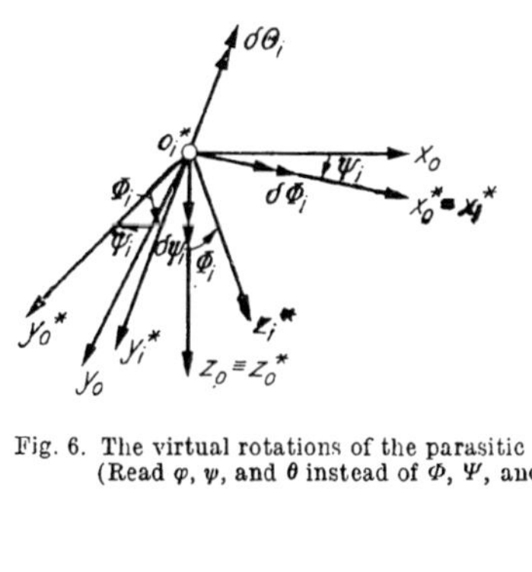

Fig. 6. The virtual rotations of the parasitic motion (Read φ, ψ, and θ instead of Φ, Ψ, and Θ)

The vector of the total virtual rotation of the wheelset with respect to $(o_i^*, x_i^*, y_i^*, z_i^*)$ is equal to

$$\begin{pmatrix}\delta\theta_{xi}^*\\ \delta\theta_{yi}^*\\ \delta\theta_{zi}^*\end{pmatrix} = \tilde{\boldsymbol{G}}_i\begin{pmatrix}\delta\theta_{0xi}^f\\ \delta\theta_{0yi}^f\\ \delta\theta_{0zi}^f\end{pmatrix} + \begin{pmatrix}\delta\theta_{xi}^{p*}\\ \delta\theta_{yi}^{p*}\\ \delta\theta_{zi}^{p*}\end{pmatrix}, \tag{4-8}$$

so that, owing to (2-11):

$$\left.\begin{aligned}\delta\theta_{xi}^* &= r_0^{-1}\sin\varphi_0\sin\psi_i\cdot\delta s_i + \delta\varphi_i\,,\\ \delta\theta_{yi}^* &= r_0^{-1}(\sin\varphi_0\cos\varphi_i\cos\psi_i + \cos\varphi_0\sin\varphi_i)\,\delta s_i - \delta\theta_i + \sin\varphi_i\cdot\delta\psi_i\,,\\ \delta\theta_{zi}^* &= r_0^{-1}(-\sin\varphi_0\sin\varphi_i\cos\psi_i + \cos\varphi_0\cos\varphi_i)\,\delta s_i + \cos\varphi_i\cdot\delta\psi_i\,.\end{aligned}\right\} \tag{4-9a}$$

The translational and angular velocities will be derived (in section 5) from (5) and (9a) respectively.

We shall also need the virtual rotation $\delta\boldsymbol{\theta}_{i0}^*$ of the coordinate system $(o_i^*, x_i^*, y_i^*, z_i^*)$: this can be found from (9a) by omitting the rotation $\delta\theta_i$ of the wheelset:

$$\left.\begin{aligned}\delta\theta_{xi0}^* &= r_0^{-1}\sin\varphi_0\sin\psi_i\cdot\delta s_i + \delta\varphi_i\,,\\ \delta\theta_{yi0}^* &= r_0^{-1}(\sin\varphi_0\cos\varphi_i\cos\psi_i + \cos\varphi_0\sin\varphi_i)\,\delta s_i + \sin\varphi_i\cdot\delta\psi_i\,,\\ \delta\theta_{zi0}^* &= r_0^{-1}(-\sin\varphi_0\sin\varphi_i\cos\psi_i + \cos\varphi_0\cos\varphi_i)\,\delta s_i + \cos\varphi_i\cdot\delta\psi_i\,.\end{aligned}\right\} \tag{4-9b}$$

With respect to the system (o_i, x_i, y_i, z_i) the components of the virtual rotation of the wheelset are determined by (2-10) and

$$\delta\boldsymbol{\theta}_i = \boldsymbol{G}_i\,\delta\boldsymbol{\theta}_{i}^*\,, \tag{4-10}$$

viz.

$$\left.\begin{aligned}\delta\theta_{xi} &= \cos\psi_i\cdot\delta\varphi_i + \cos\varphi_i\sin\psi_i\cdot\delta\theta_i\,,\\ \delta\theta_{yi} &= r_0^{-1}\sin\varphi_0\cdot\delta s_i + \sin\psi_i\cdot\delta\varphi_i - \cos\varphi_i\cos\psi_i\cdot\delta\theta_i\,,\\ \delta\theta_{zi} &= r_0^{-1}\cos\varphi_0\cdot\delta s_i - \sin\varphi_i\cdot\delta\theta_i + \delta\psi_i\,.\end{aligned}\right\} \tag{4-11}$$

We now shall calculate the virtual displacement of a wheel in the contact point. According to (2-12) we have

$$\boldsymbol{\varrho}_i = \boldsymbol{\varrho}_{0i} + \boldsymbol{G}_{0i}\boldsymbol{q}_i + \boldsymbol{G}_{0i}\boldsymbol{G}_i\boldsymbol{r}_i^*\,. \tag{4-12}$$

The virtual displacement of the wheel with components opposite to the directions $o_i^*x_i^*$, $o_i^*y_i^*$, $o_i^*z_i^*$ are

$$\delta\boldsymbol{s}_{ij}^* = -\tilde{\boldsymbol{G}}_i\tilde{\boldsymbol{G}}_{0i}\,\delta\boldsymbol{\varrho}_i \tag{4-13}$$

putting $\boldsymbol{r}_i^*$ equal to the radius vector of the contact point we find from (12)

$$\delta\boldsymbol{s}_{ij}^* = -\tilde{\boldsymbol{G}}_i(\tilde{\boldsymbol{G}}_{0i}\,\delta\boldsymbol{\varrho}_{0i} + \tilde{\boldsymbol{G}}_{0i}\,\delta\boldsymbol{G}_{0i}\cdot\boldsymbol{q}_i + \delta\boldsymbol{q}_i) - (\tilde{\boldsymbol{G}}_i\tilde{\boldsymbol{G}}_{0i}\,\delta\boldsymbol{G}_{0i}\cdot\boldsymbol{G}_i + \tilde{\boldsymbol{G}}_i\,\delta\boldsymbol{G}_i)\,\boldsymbol{r}_i^*\,. \tag{4-14}$$

The vector

$$\tilde{\boldsymbol{G}}_{0i}\,\delta\boldsymbol{\varrho}_{0i} + \tilde{\boldsymbol{G}}_{0i}\,\delta\boldsymbol{G}_{0i}\cdot\boldsymbol{q}_i + \delta\boldsymbol{q}_i$$

is, according to (2), equal to $\delta\boldsymbol{r}_{i0}$ (4) and its coordinates are given by (5). Moreover, we can write

$$\tilde{\boldsymbol{G}}_i\tilde{\boldsymbol{G}}_{0i}\,\delta\boldsymbol{G}_{0i}\cdot\boldsymbol{G}_i + \tilde{\boldsymbol{G}}_i\,\delta\boldsymbol{G}_i = \delta\bar{\boldsymbol{\Theta}}_i^*\,, \tag{4-15}$$

where

$$\delta\bar{\boldsymbol{\Theta}}_i^* = \begin{pmatrix}0 & -\delta\theta_{zi}^* & \delta\theta_{yi}^*\\ \delta\theta_{zi}^* & 0 & -\delta\theta_{xi}^*\\ -\delta\theta_{yi}^* & \delta\theta_{xi}^* & 0\end{pmatrix} \tag{4-16}$$

and the coordinates of $\delta\bar{\boldsymbol{\Theta}}_i^*$ are given by (9a). For the vector $\boldsymbol{r}_i^*$ we have

$$\tilde{\boldsymbol{r}}_i^* = (x_{ij}^*, y_{ij}^*, z_{ij}^*)\,. \tag{4-17}$$

Altogether we find for the virtual displacement in the contact point, using (3-29a—c):

$$\begin{aligned}\delta s_{xij}^* = \{&-\cos\psi_i + r_0^{-1}(v_i\cos\varphi_0 - w_i\sin\varphi_0)\cos\psi_i - r_0^{-1}b_{yzij}\sin\varphi_0\cos\psi_i + r_0^{-1}a_{yzij}\cos\varphi_0\}\,\delta s_i -\\ &- \sin\psi_{ij}\,\delta v_i + z_{ij}^*\,\delta\theta_i + a_{yzij}\,\delta\psi_i\,,\end{aligned} \tag{4-18a}$$

$$\begin{aligned}\delta s_{yij}^* = \{&\cos\varphi_i\sin\psi_i - r_0^{-1}(v_i\cos\varphi_0 - w_i\sin\varphi_0)\cos\varphi_i\sin\psi_i + r_0^{-1}(a_{xzij}\sin\varphi_0 - x_{ij}^*\cos\varphi_0\cos\varphi_i)\}\,\delta s_i -\\ &- \cos\varphi_i\cos\psi_i\cdot\delta v_i - \sin\varphi_i\cdot\delta w_i + z_{ij}^*\,\delta\varphi_i - x_{ij}^*\cos\varphi_i\cdot\delta\psi_i\,,\end{aligned} \tag{4-18b}$$

$$\begin{aligned}\delta s_{zij}^* = \{&-\sin\varphi_i\sin\psi_i + r_0^{-1}(v_i\cos\varphi_0 - w_i\sin\varphi_0)\sin\varphi_i\sin\psi_i + r_0^{-1}(a_{xyij}\sin\varphi_0 + x_{ij}^*\cos\varphi_0\sin\varphi_i)\}\,\delta s_i +\\ &+ \sin\varphi_i\cos\psi_i\cdot\delta v_i - \cos\varphi_i\cdot\delta w_i - y_{ij}^*\,\delta\varphi_i - x_{ij}^*\,\delta\theta_i + x_{ij}^*\sin\varphi_i\cdot\delta\psi_i\,.\end{aligned} \tag{4-18c}$$

The components of the virtual displacement in the contact point into the normal and the tangential directions are found by means of the relation

$$\begin{pmatrix}\delta s_{nij}\\ \delta s_{txij}\\ \delta s_{tyij}\end{pmatrix} = \tilde{\boldsymbol{G}}_{ij}\begin{pmatrix}\delta s_{xij}^*\\ \delta s_{yij}^*\\ \delta s_{zij}^*\end{pmatrix} \tag{4-19}$$

with $\widetilde{\boldsymbol{G}}_{ij}$ (3-32). Now we can neglect all terms with r_0^{-2}. Then we find, using (3-28):

$$\begin{aligned}\delta s_{nij} = {} & \mp \sin\gamma_{ij}\cdot\delta v_i - \cos\gamma_{ij}\cdot\delta w_i + (\pm\, b_{yzij}\sin\gamma_{ij}\cos\psi_i - a_{yzij}\cos\gamma_{ij})\,\delta\varphi_i \pm \\ & \pm(-x_{ij}^*\cos\psi_i + a_{yzij}\sin\psi_i)\sin\gamma_{ij}\cdot\delta\psi_i + \\ & + r_0^{-1}x_{ij}\{\pm\, b_{yzij}\sin\psi_i\cdot\delta\varphi_i - (x_{ij}^*\sin\psi_i + a_{yzij}\cos\psi_i)\,\delta\psi_i\}\sin\beta_{ij}\,, \end{aligned} \tag{4-20a}$$

$$\begin{aligned}\delta s_{txij} = {} & -\{1 - r_0^{-1}(v_i\cos\varphi_0 - w_i\sin\varphi_0) - r_0^{-1}x_{ij}^*\cos\varphi_0\sin\psi_i - r_0^{-1}a_{yzij}\cos\varphi_0\cos\psi_i + \\ & + r_0^{-1}b_{yzij}\sin\varphi_0\}\,\delta s_i - b_{yzij}\sin\psi_i\cdot\delta\varphi_i + b_{xzij}\,\delta\theta_i + (x_{ij}^*\sin\psi_i + a_{yzij}\cos\psi_i)\,\delta\psi_i + \\ & + r_0^{-1}x_{ij}\,\{-\cos\varphi_0\cdot\delta v_i + \sin\varphi_0\cdot\delta w_i + (b_{yzij}\cos\varphi_0\cos\psi_i + a_{yzij}\sin\varphi_0)\,\delta\varphi_i + \\ & + (a_{xzij}\cos\varphi_0 + x_{ij}^*\sin\varphi_0\cos\varphi_i)\,\delta\theta_i + (-x_{ij}^*\cos\psi_i + a_{yzij}\sin\psi_i)\cos\varphi_0\cdot\delta\psi_i\}\,, \end{aligned} \tag{4-20b}$$

$$\begin{aligned}\delta s_{tyij} = {} & -\cos\gamma_{ij}\cdot\delta v_i \pm \sin\gamma_{ij}\cdot\delta w_i + (b_{yzij}\cos\gamma_{ij}\cos\psi_i \pm a_{yzij}\sin\gamma_{ij})\,\delta\varphi_i + \\ & + (a_{xzij}\cos\gamma_{ij} \pm x_{ij}^*\sin\gamma_{ij}\cos\varphi_i)\,\delta\theta_i + (-x_{ij}^*\cos\psi_i + a_{yzij}\sin\psi_i)\cos\gamma_{ij}\cdot\delta\psi_i + \\ & + r_0^{-1}x_{ij}\,\{b_{yzij}\sin\psi_i\cdot\delta\varphi_i - b_{xzij}\,\delta\theta_i - (x_{ij}^*\sin\psi_i + a_{yzij}\cos\psi_i)\,\delta\psi_i\}\cos\beta_{ij}\,. \end{aligned} \tag{4-20c}$$

The normal displacements δs_{nij} have to be zero, which might be proved by varying all the quantities $x_{ij}, \dots, z_{ij}^*$, $r_{ij}^*, \gamma_{ij}, \gamma_{ij}^*, a_{xyij}, \dots, b_{yzij}$ and the coordinates v_i, ψ_i, w_i and φ_i in (3-13), (3-19), (3-28), (3-29a—c) and (3-31) and by solving δw_i and $\delta\varphi_i$ as linear functions of δv_i and $\delta\psi_i$; inserting these δw_i and $\delta\varphi_i$ into (20a) we shall find $\delta s_{nij} = 0$.

Assuming δs_{nij} to be zero we can find from (20a) two relations between the variations δw_i, $\delta\varphi_i$, δv_i and $\delta\psi_i$. Solving for δw_i and $\delta\varphi_i$ we find the four partial derivatives

$$\frac{\partial w_i}{\partial v_i}, \quad \frac{\partial w_i}{\partial \psi_i}, \quad \frac{\partial \varphi_i}{\partial v_i} \quad \text{and} \quad \frac{\partial \varphi_i}{\partial \psi_i},$$

which enable us to draw up the constraint relations

$$w_i = w_i(v_i, \psi_i)\,, \qquad \varphi_i = \varphi_i(v_i, \psi_i)\,. \tag{4-21}$$

5. Kinematics

The velocity of the wheelset is determined by the translational velocity $\boldsymbol{V}_i = \dot{\boldsymbol{r}}_{i0}$ of the mass centre and the angular velocity $\boldsymbol{\omega}_i^*$. By means of (4-5) we find for the components of the mass centre velocity in the directions of (o_i, x_i, y_i, z_i):

$$V_{xi} = \dot{s}_i + r_0^{-1}(-v_i\cos\varphi_0 + w_i\sin\varphi_0)\,\dot{s}_i\,, \qquad V_{yi} = \dot{v}_i\,, \qquad V_{zi} = \dot{w}_i\,. \tag{5-1}$$

Moreover,

$$\boldsymbol{\omega}_i^* = \dot{\boldsymbol{\theta}}_i^*\,, \tag{5-2}$$

so that the components of the angular velocity in the directions of $(o_i^*, x_i^*, y_i^*, z_i^*)$ are, according to (4.9a):

$$\left.\begin{aligned} \omega_{xi}^* &= r_0^{-1}\dot{s}_i\sin\varphi_0\sin\psi_i + \dot{\varphi}_i\,, \\ \omega_{yi}^* &= r_0^{-1}\dot{s}_i(\sin\varphi_0\cos\varphi_i\cos\psi_i + \cos\varphi_0\sin\varphi_i) - \dot{\theta}_i + \dot{\psi}_i\sin\varphi_i\,, \\ \omega_{zi}^* &= r_0^{-1}\dot{s}_i(-\sin\varphi_0\sin\varphi_i\cos\psi_i + \cos\varphi_0\cos\varphi_i) + \dot{\psi}_i\cos\varphi_i\,. \end{aligned}\right\} \tag{5-3}$$

The angular velocity of the coordinate system $(o_i^*, x_i^*, y_i^*, z_i^*)$ can be derived from the virtual rotation $\delta\boldsymbol{\theta}_{0i}^*$:

$$\boldsymbol{\omega}_{i0}^* = \dot{\boldsymbol{\theta}}_{i0}^*\,, \tag{5-4}$$

so that because of (4-9b):

$$\left.\begin{aligned} \omega_{xi0}^* &= r_0^{-1}\dot{s}_i\sin\varphi_0\sin\psi_i + \dot{\varphi}_i\,, \\ \omega_{yi0}^* &= r_0^{-1}\dot{s}_i(\sin\varphi_0\cos\varphi_i\cos\psi_i + \cos\varphi_0\sin\varphi_i) + \dot{\psi}_i\sin\varphi_i\,, \\ \omega_{zi0}^* &= r_0^{-1}\dot{s}_i(-\sin\varphi_0\sin\varphi_i\cos\psi_i + \cos\varphi_0\cos\varphi_i) + \dot{\psi}_i\cos\varphi_i\,. \end{aligned}\right\} \tag{5-5}$$

The angular velocity of the wheelset with components related to the system (o_i, x_i, y_i, z_i) is equal to $\boldsymbol{\omega}_i = \dot{\boldsymbol{\theta}}_i$, so that in connection with (4-11):

$$\left.\begin{aligned} \omega_{xi} &= \dot{\varphi}_i\cos\psi_i + \dot{\theta}_i\cos\varphi_i\sin\psi_i\,, \\ \omega_{yi} &= \dot{s}_i r_0^{-1}\sin\varphi_0 + \dot{\varphi}_i\sin\psi_i - \dot{\theta}_i\cos\varphi_i\cos\psi_i\,, \\ \omega_{zi} &= \dot{s}_i r_0^{-1}\cos\varphi_0 - \dot{\theta}_i\sin\varphi_i + \dot{\psi}_i\,. \end{aligned}\right\} \tag{5-6}$$

From this we easily can derive the components of the angular velocity $\boldsymbol{\omega}_{i0}$ of the system (o_i, x_i, y_i, z_i) itself:

$$\omega_{xi0} = 0\,, \qquad \omega_{yi0} = r_0^{-1}\dot{s}_i\sin\varphi_0\,, \qquad \omega_{zi0} = r_0^{-1}\dot{s}_i\cos\varphi_0\,. \tag{5-7}$$

The velocity of a wheel in a momentary contact point is determined by the relations

$$W_{nij} = -\dot{s}_{nij}\,, \qquad W_{txij} = -\dot{s}_{txij}\,, \qquad W_{tyij} = -\dot{s}_{tyij}\,. \tag{5-8}$$

when the components are taken into the positive directions n_{ij}, t_{xij}, t_{yij}. Owing to (4-20 a—c) we have

$$W_{nij} = \pm \dot{v}_i \sin\gamma_{ij} + \dot{w}_i \cos\gamma_{ij} + \dot{\varphi}_i(\mp b_{yzij} \sin\gamma_{ij} \cos\psi_i + a_{yzij}\cos\gamma_{ij}) \pm$$
$$\pm \dot{\psi}_i(x^*_{ij}\cos\psi_i - a_{yzij}\sin\psi_i)\sin\gamma_{ij} + r_0^{-1}x_{ij}\{\mp\dot{\varphi}_i b_{yzij}\sin\psi_i + \dot{\psi}_i(x^*_{ij}\sin\psi_i + a_{yzij}\cos\psi_i)\}\sin\beta_{ij}\,, \tag{5-9a}$$

$$W_{txij} = \dot{s}_i\{1 - r_0^{-1}(v_i\cos\varphi_0 - w_i\sin\varphi_0) - r_0^{-1}x^*_{ij}\cos\varphi_0\sin\psi_i - r_0^{-1}a_{yzij}\cos\varphi_0\cos\psi_i +$$
$$+ r_0^{-1}b_{yzij}\sin\varphi_0\} + \dot{\varphi}_i b_{yzij}\sin\psi_i - \dot{\theta}_i b_{xzij} - \dot{\psi}_i(x^*_{ij}\sin\psi_i + a_{yzij}\cos\psi_i) +$$
$$+ r_0^{-1}x_{ij}\{\dot{v}_i\cos\varphi_0 - \dot{w}_i\sin\varphi_0 - \dot{\varphi}_i(b_{yzij}\cos\varphi_0\cos\psi_i + a_{yzij}\sin\varphi_0) -$$
$$- \dot{\theta}_i(a_{xzij}\cos\varphi_0 + x^*_{ij}\sin\varphi_0\cos\varphi_i) + \dot{\psi}_i(x^*_{ij}\cos\psi_i - a_{yzij}\sin\psi_i)\cos\varphi_0\}\,, \tag{5-9b}$$

$$W_{tyij} = \dot{v}_i\cos\gamma_{ij} \mp \dot{w}_i\sin\gamma_{ij} - \dot{\varphi}_i(b_{yzij}\cos\gamma_{ij}\cos\psi_i \pm a_{yzij}\sin\gamma_{ij}) -$$
$$- \dot{\theta}_i(a_{xzij}\cos\gamma_{ij} \pm x^*_{ij}\sin\gamma_{ij}\cos\varphi_i) + \dot{\psi}_i(x^*_{ij}\cos\psi_i - a_{yzij}\sin\psi_i)\cos\gamma_{ij} +$$
$$+ r_0^{-1}x_{ij}\{-\dot{\varphi}_i b_{yzij}\sin\psi_i + \dot{\theta}_i b_{xzij} + \dot{\psi}_i(x^*_{ij}\sin\psi_i + a_{yzij}\cos\psi_i)\}\cos\beta_{ij}\,. \tag{5-9c}$$

For $r = \infty$ these formulae agree with the corresponding formulae (2-1)—(2-2b) for the motion on a tangent track [5 p. 400].

Because of the constraint relations the normal velocities W_{nij} have to be zero. The tangential velocities W_{txij} and W_{tyij} will be needed for determining the tangential forces, for which we also have to know the component of the angular velocity ω_{nij} of each wheel about the direction of the normal in the contact point n_{ij}. We can write

$$\begin{pmatrix}\omega_{nij}\\ \omega_{txij}\\ \omega_{tyij}\end{pmatrix} = \tilde{G}_{ij}\boldsymbol{\omega}^*_i\,, \tag{5-10}$$

which, together with (3-32) and (3), yields

$$\omega_{nij} = r_0^{-1}\dot{s}_i\cos\beta_{ij} \pm \dot{\varphi}_i\sin\gamma_{ij}\sin\psi_i - \dot{\theta}_i(\pm\sin\gamma_{ij}\cos\varphi_i\cos\psi_i + \cos\gamma_{ij}\sin\varphi_i) + \dot{\psi}_i\cos\gamma_{ij} \mp$$
$$\mp r_0^{-1}x_{ij}(\dot{\varphi}_i\cos\psi_i + \dot{\theta}_i\cos\varphi_i\sin\psi_i)\sin\beta_{ij}\,. \tag{5-11}$$

6. Forces and work coefficients

In the investigation on the motion along a tangent track [5, 6] we assumed that the wheelset was loaded by the force of gravity $m_i g$, applied at the mass centre, and by two constant spring forces. In the present investigation we will make a more general assumption and we shall suppose that the wheelset is loaded by an external system of forces with a resultant $\boldsymbol{F}_i$ in the mass centre o^*_i and a resultant moment $\boldsymbol{M}_i$:

$$\tilde{\boldsymbol{F}}_i = (F_{xi}, F_{yi}, F_{zi})\,, \qquad \tilde{\boldsymbol{M}}_i = (M_{xi}, M_{yi}, M_{zi})\,, \tag{6-1}$$

the components being taken into the directions of the axes of the system (o_i, x_i, y_i, z_i). For these forces we have to determine the work coefficients. The virtual work δW_{ex} of $\boldsymbol{F}_i$ and $\boldsymbol{M}_i$ is equal to

$$\delta W_{ex} = \boldsymbol{F}_i\cdot\delta\boldsymbol{r}_{i0} + \boldsymbol{M}_i\cdot\delta\boldsymbol{\theta}_i\,, \tag{6-2}$$

$\delta\boldsymbol{r}_{i0}$ being determined by (4-5) and $\delta\boldsymbol{\theta}_i$ by (4-11). The potential energy of the system is $U = 0$.

In a contact point the normal force N_{ij} and the tangential forces T_{xij} and T_{yij} are applied at the wheel. They are taken negative into the directions n_{ij}, t_{xij}, t_{yij} respectively. The work coefficients of these forces ensue from the expression for their virtual work

$$\delta W_{in} = \sum_{j=1}^{2}(N_{ij}\,\delta s_{nij} + T_{xij}\,\delta s_{txij} + T_{yij}\,\delta s_{tyij})\,. \tag{6-3}$$

When the constraint relations (4-21) hold, the virtual work of the normal forces disappears. For the time being we shall not take into account these relations, so that the work coefficients will also depend on the normal forces.

The total virtual work

$$\delta W = \delta W_{ex} + \delta W_{in} \tag{6-4}$$

is equivalent with

$$\delta W = Q_{si}\,\delta s_i + Q_{vi}\,\delta v_i + Q_{wi}\,\delta w_i + Q_{\varphi i}\,\delta\varphi_i + Q_{\theta i}\,\delta\theta_i + Q_{\psi i}\,\delta\psi_i\,. \tag{6-5}$$

Substituting (4-20 a—c) into (3), using (2), (4-5), (4-11) and comparing the coefficients of the variations $\delta s_i, \ldots, \delta\psi_i$ yields

$$Q_{si} = F_{xi}\{1 - r_0^{-1}(v_i\cos\varphi_0 - w_i\sin\varphi_0)\} + r_0^{-1}(M_{yi}\sin\varphi_0 + M_{zi}\cos\varphi_0) -$$
$$- \sum_{j=1}^{2} T_{xij}\{1 - r_0^{-1}(v_i\cos\varphi_0 - w_i\sin\varphi_0) - r_0^{-1}(x^*_{ij}\sin\psi_i + a_{yzij}\cos\psi_i)\cos\varphi_0 + r_0^{-1}b_{yzij}\sin\varphi_0\}\,, \tag{6-6a}$$

$$Q_{vi} = F_{yi} + \sum_{j=1}^{2}(-1)^j N_{ij}\sin\gamma_{ij} - \sum_{j=1}^{2} T_{yij}\cos\gamma_{ij} - r_0^{-1}\cos\varphi_0\sum_{j=1}^{2} T_{xij}x_{ij}\,, \tag{6-6b}$$

$$Q_{wi} = F_{zi} - \sum_{j=1}^{2} N_{ij}\cos\gamma_{ij} - \sum_{j=1}^{2}(-1)^j T_{yij}\sin\gamma_{ij} + r_0^{-1}\sin\varphi_0\sum_{j=1}^{2} T_{xij}x_{ij}\,, \tag{6-6c}$$

$$Q_{\varphi i} = M_{xi} \cos \psi_i + M_{yi} \sin \psi_i - \sum_{j=1}^{2} N_{ij}\{(-1)^j b_{yzij} \sin \gamma_{ij} \cos \psi_i + a_{yzij} \cos \gamma_{ij}\} -$$
$$- \sum_{j=1}^{2} T_{xij} b_{yzij} \sin \psi_i + \sum_{j=1}^{2} T_{yij}\{b_{yzij} \cos \gamma_{ij} \cos \psi_i - (-1)^j a_{yzij} \sin \gamma_{ij}\} -$$
$$- r_0^{-1} \sum_{j=1}^{2} (-1)^j N_{ij} x_{ij} b_{yzij} \sin \psi_i \sin \beta_{ij} + r_0^{-1} \sum_{j=1}^{2} T_{xij} x_{ij} (b_{yzij} \cos \varphi_0 \cos \psi_i + a_{yzij} \sin \varphi_0) +$$
$$+ r_0^{-1} \sum_{j=1}^{2} T_{yij} x_{ij} b_{yzij} \sin \psi_i \cos \beta_{ij} , \tag{6-7a}$$

$$Q_{\theta i} = (M_{xi} \sin \psi_i - M_{yi} \cos \psi_i) \cos \varphi_i - M_{zi} \sin \varphi_i + \sum_{j=1}^{2} T_{xij} b_{xzij} +$$
$$+ \sum_{j=1}^{2} T_{yij}\{a_{xzij} \cos \gamma_{ij} - (-1)^j x_{ij}^* \sin \gamma_{ij} \cos \varphi_i\} + r_0^{-1} \sum_{j=1}^{2} T_{xij} x_{ij} (a_{xzij} \cos \varphi_0 + x_{ij}^* \sin \varphi_0 \cos \varphi_i) -$$
$$- r_0^{-1} \sum_{j=1}^{2} T_{yij} x_{ij} b_{xzij} \cos \beta_{ij} , \tag{6-7b}$$

$$Q_{\psi i} = M_{zi} + \sum_{j=1}^{2} (-1)^j N_{ij} (x_{ij}^* \cos \psi_i - a_{xzij} \sin \psi_i) \sin \gamma_{ij} + \sum_{j=1}^{2} T_{xij} (x_i^* \sin \psi_i + a_{yzij} \cos \psi_i) -$$
$$- \sum_{j=1}^{2} T_{yij} (x_{ij}^* \cos \psi_i - a_{yzij} \sin \psi_i) \cos \gamma_{ij} + r_0^{-1} \sum_{j=1}^{2} (-1)^j N_{ij} x_{ij} (x_{ij}^* \sin \psi_i + a_{yzij} \cos \psi_i) \sin \beta_{ij} -$$
$$- r_0^{-1} \sum_{j=1}^{2} T_{xij} x_{ij} (x_{ij}^* \cos \psi_i - a_{yzij} \sin \psi_i) \cos \varphi_0 - r_0^{-1} \sum_{j=1}^{2} T_{yij} x_{ij} (x_{ij}^* \sin \psi_i + a_{yzij} \cos \psi_i) \cos \beta_{ij} . \tag{6-7c}$$

For $r_0 = \infty$ these formulae agree with the formulae (2-8a)—(2-9c) for the motion on a tangent track [5 p. 401—402].

7. Kinetic energy

The moments of inertia of a wheelset with respect to the axes $o_i^* x_i^*$, $o_i^* y_i^*$, $o_i^* z_i^*$ are indicated by J_i, J_{yi} and J_i respectively. Then the kinetic energy is equal to

$$T = \tfrac{1}{2} m_i (V_{xi}^2 + V_{yi}^2 + V_{zi}^2) + \tfrac{1}{2} (J_i \omega_{xi}^{*2} + J_{yi} \omega_{yi}^{*2} + J_i \omega_{zi}^{*2}) . \tag{7-1}$$

By means of (5-1) and (5-3) and by omitting all terms with r_0^{-2} we find

$$T = \tfrac{1}{2} m_i (\dot{s}_i^2 + \dot{v}_i^2 + \dot{w}_i^2) + \tfrac{1}{2} J_i (\dot{\varphi}_i^2 + \dot{\psi}_i^2 \cos^2 \varphi_i) + \tfrac{1}{2} J_{yi} (\theta_i - \dot{\psi}_i \sin \varphi_i)^2 +$$
$$+ m_i r_0^{-1} (-v_i \cos \varphi_0 + w_i \sin \varphi_0) \dot{s}_i^2 +$$
$$+ J_i r_0^{-1} \dot{s}_i \{\dot{\varphi}_i \sin \varphi_0 \sin \psi_i + \dot{\psi}_i (-\sin \varphi_0 \sin \varphi_i \cos \psi_i + \cos \varphi_0 \cos \varphi_i) \cos \varphi_i\} +$$
$$+ J_{yi} r_0^{-1} \dot{s}_i (\sin \varphi_0 \cos \varphi_i \cos \psi_i + \cos \varphi_0 \sin \varphi_i) (-\dot{\theta}_i + \dot{\psi}_i \sin \varphi_i) . \tag{7-2}$$

For $r_0 = \infty$ this expression is identical with the expression (2-5) for the motion on a tangent track [5 p. 401].

8. The equations of motion

The potential energy of the system being zero, LAGRANGE's equations read

$$\frac{d}{dt} \frac{\partial T}{\partial \dot{q}_k} - \frac{\partial T}{\partial q_k} = Q_k \qquad (k = 1, \ldots, 6; q_1 = s_i, \ldots, q_6 = \psi_i) . \tag{8-1}$$

By means of (7-2) we obtain

$$\frac{d}{dt} [m_i \dot{s}_i + 2 m_i r_0^{-1} \dot{s}_i (-v_i \cos \varphi_0 + w_i \sin \varphi_0) +$$
$$+ J_i r_0^{-1} \{\dot{\varphi}_i \sin \varphi_0 \sin \psi_i + \dot{\psi}_i (-\sin \varphi_0 \sin \varphi_i \cos \psi_i + \cos \varphi_0 \cos \varphi_i) \cos \varphi_i\} +$$
$$+ J_{yi} r_0^{-1} (\sin \varphi_0 \cos \varphi_i \cos \psi_i + \cos \varphi_0 \sin \varphi_i) (-\dot{\theta}_i + \dot{\psi}_i \sin \varphi_i)] = Q_{si} , \tag{8-2a}$$

$$\frac{d}{dt} (m_i \dot{v}_i) + m_i r_0^{-1} \dot{s}_i^2 \cos \varphi_0 = Q_{vi} , \tag{8-2b}$$

$$\frac{d}{dt} (m_i \dot{w}_i) - m_i r_0^{-1} \dot{s}_i^2 \sin \varphi_0 = Q_{wi} , \tag{8-2c}$$

$$\frac{d}{dt} (J_i \dot{\varphi}_i + J_i r_0^{-1} \dot{s}_i \sin \varphi_0 \sin \psi_i) + J_i \dot{\psi}_i^2 \cos \varphi_i \sin \varphi_i + J_{yi} (\dot{\theta}_i - \dot{\psi}_i \sin \varphi_i) \dot{\psi}_i \cos \varphi_i +$$
$$+ (J_i - J_{yi}) r_0^{-1} \dot{s}_i \dot{\psi}_i (\sin \varphi_0 \cos 2\varphi_i \cos \psi_i + \cos \varphi_0 \sin 2\varphi_i) -$$
$$- J_{yi} r_0^{-1} \dot{s}_i \dot{\theta}_i (\sin \varphi_0 \sin \varphi_i \cos \psi_i - \cos \varphi_0 \cos \varphi_i) = Q_{\varphi i} , \tag{8-3a}$$

$$\frac{d}{dt} J_{yi}\{\dot{\theta}_i - \dot{\psi}_i \sin\varphi_i - r_0^{-1}\dot{s}_i(\sin\varphi_0\cos\varphi_i\cos\psi_i + \cos\varphi_0\sin\varphi_i)\} = Q_{\theta i}\,, \tag{8-3b}$$

$$\frac{d}{dt}\{J_i\dot{\psi}_i\cos^2\varphi_i - J_{yi}(\dot{\theta}_i - \dot{\psi}_i\sin\varphi_i)\sin\varphi_i - (J_i - J_{yi})\, r_0^{-1}\dot{s}_i\sin\varphi_0\cos\varphi_i\sin\varphi_i\cos\psi_i + $$
$$+ (J_i\cos^2\varphi_i + J_{yi}\sin^2\varphi_i)\, r_0^{-1}\dot{s}_i\cos\varphi_0\} - J_i r_0^{-1}\dot{s}_i(\dot{\varphi}_i\cos\psi_i + \dot{\psi}_i\sin\varphi_i\cos\varphi_i\sin\psi_i)\sin\varphi_0 -$$
$$- J_{yi} r_0^{-1}\dot{s}_i(\dot{\theta}_i - \dot{\psi}_i\sin\varphi_i)\sin\varphi_0\cos\varphi_i\sin\psi_i = Q_{\psi i}\,. \tag{8-3c}$$

The generalized forces are given by (6-6a)–(6-7c).

Evaluating the equations (2a)–(3c) yields

$$m_i\ddot{s}_i + 2m_i r_0^{-1}\ddot{s}_i(-v_i\cos\varphi_0 + w_i\sin\varphi_0) + 2m_i r_0^{-1}\dot{s}_i(-\dot{v}_i\cos\varphi_0 + \dot{w}_i\sin\varphi_0) +$$
$$+ J_i r_0^{-1}\{\ddot{\varphi}_i\sin\varphi_0\sin\psi_i + \ddot{\psi}_i(-\sin\varphi_0\sin\varphi_i\cos\psi_i + \cos\varphi_0\cos\varphi_i)\cos\varphi_i\} +$$
$$+ J_i r_0^{-1}\dot{\psi}_i\{2\dot{\varphi}_i(\sin\varphi_0\sin\varphi_i\cos\psi_i - \cos\varphi_0\cos\varphi_i) + \dot{\psi}_i\sin\varphi_0\cos\varphi_i\sin\psi_i\}\sin\varphi_i +$$
$$+ J_{yi} r_0^{-1}(\dot{\theta}_i - \dot{\psi}_i\sin\varphi_i)(\dot{\varphi}_i\sin\varphi_0\sin\varphi_i\cos\psi_i - \dot{\varphi}_i\cos\varphi_0\cos\varphi_i + \dot{\psi}_i\sin\varphi_0\cos\varphi_i\sin\psi_i) +$$
$$+ J_{yi} r_0^{-1}(-\ddot{\theta}_i + \ddot{\psi}_i\sin\varphi_i + \dot{\varphi}_i\dot{\psi}_i\cos\varphi_i)(\sin\varphi_0\cos\varphi_i\cos\psi_i + \cos\varphi_0\sin\varphi_i) = Q_{si}\,, \tag{8-4a}$$

$$m_i\ddot{v}_i + m_i r_0^{-1}\dot{s}_i^2\cos\varphi_0 = Q_{vi}\,, \tag{8-4b}$$

$$m_i\ddot{w}_i - m_i r_0^{-1}\dot{s}_i^2\sin\varphi_0 = Q_{wi}\,, \tag{8-4c}$$

$$J_i\ddot{\varphi}_i + J_i\dot{\psi}_i^2\cos\varphi_i\sin\varphi_i + J_{yi}(\dot{\theta}_i - \dot{\psi}_i\sin\varphi_i)\dot{\psi}_i\cos\varphi_i +$$
$$+ J_i r_0^{-1}\ddot{s}_i\sin\varphi_0\sin\psi_i + J_i r_0^{-1}\dot{s}_i\dot{\psi}_i(2\sin\varphi_0\cos^2\varphi_i\cos\psi_i + \cos\varphi_0\sin 2\varphi_i) -$$
$$- J_{yi} r_0^{-1}\dot{s}_i\{\dot{\theta}_i(\sin\varphi_0\sin\varphi_i\cos\psi_i - \cos\varphi_0\cos\varphi_i) + \dot{\psi}_i(\sin\varphi_0\cos 2\varphi_i\cos\psi_i + \cos\varphi_0\sin 2\varphi_i)\} = Q_{\varphi i}\,, \tag{8-5a}$$

$$J_{yi}(\ddot{\theta}_i - \ddot{\psi}_i\sin\varphi_i - \dot{\varphi}_i\dot{\psi}_i\cos\varphi_i) - J_{yi} r_0^{-1}\ddot{s}_i(\sin\varphi_0\cos\varphi_i\cos\psi_i + \cos\varphi_0\sin\varphi_i) +$$
$$+ J_{yi} r_0^{-1}\dot{s}_i(\dot{\varphi}_i\sin\varphi_0\sin\varphi_i\cos\psi_i - \dot{\varphi}_i\cos\varphi_0\cos\varphi_i + \dot{\psi}_i\sin\varphi_0\cos\varphi_i\sin\psi_i) = Q_{\theta i}\,, \tag{8-5b}$$

$$J_i(\ddot{\psi}_i\cos\varphi_i - 2\dot{\varphi}_i\dot{\psi}_i\sin\varphi_i)\cos\varphi_i - J_{yi}(\ddot{\theta}_i - \ddot{\psi}_i\sin\varphi_i - 2\dot{\varphi}_i\dot{\psi}_i\cos\psi_i)\sin\varphi_i -$$
$$- J_{yi}\dot{\varphi}_i\dot{\theta}_i\cos\varphi_i + J_i r_0^{-1}\ddot{s}_i(-\sin\varphi_0\sin\varphi_i\cos\psi_i + \cos\varphi_0\cos\varphi_i)\cos\varphi_i -$$
$$- J_i r_0^{-1}\dot{s}_i\dot{\varphi}_i(2\sin\varphi_0\cos^2\varphi_i\cos\psi_i + \cos\varphi_0\sin 2\varphi_i) +$$
$$+ J_{yi} r_0^{-1}\ddot{s}_i(\sin\varphi_0\cos\varphi_i\cos\psi_i + \cos\varphi_0\sin\varphi_i)\sin\varphi_i +$$
$$+ J_{yi} r_0^{-1}\dot{s}_i\dot{\varphi}_i(\sin\varphi_0\cos 2\varphi_i\cos\psi_i + \cos\varphi_0\sin 2\varphi_i) - J_{yi} r_0^{-1}\dot{s}_i\dot{\theta}_i\sin\varphi_0\cos\varphi_i\sin\psi_i = Q_{\psi i}\,. \tag{8-5c}$$

For $r_0 = \infty$ these equations are identical with the equations of motion (2-6)–(2-7c) for a tangent track [5 p. 401].

Multiplying (4a) with $\dot{s}_i$, (4b) with $\dot{v}_i$, ... and (5c) with $\dot{\psi}_i$ and adding up, we obtain the energy balance

$$\frac{d}{dt}(T + U) = P_{ex} + P_{in}\,, \tag{8-6}$$

where T is given by (7-2), $U = 0$, whereas the power of the external forces is equal to

$$P_{ex} = F_{xi}V_{xi} + F_{yi}V_{yi} + F_{zi}V_{zi} + M_{xi}\omega_{xi} + M_{yi}\omega_{yi} + M_{zi}\omega_{zi} \tag{8-7a}$$

and the power of the internal forces to

$$P_{in} = -\sum_{j=1}^{2}(N_{ij}W_{nij} + T_{xij}W_{txij} + T_{yij}W_{tyij}); \tag{8-7b}$$

for the velocites and the angular velocities the relations (5-1), (5-6) and (5-9 a–c) hold respectively. (Because $W_{nij} = 0$, we can reduce (7b) to

$$P_{in} = -\sum_{j=1}^{2}(T_{xij}W_{txij} + T_{yij}W_{tyij}).) \tag{8-8}$$

In this way a check on the equations of motion is obtained; a second one will be shown in the next section.

9. Alternative derivation of the equations of motion

We shall now verify the equations of motion (9-4a)–(9-5c) by comparing them with the equations ensuing from NEWTON's laws. With respect to the system (o_i, x_i, y_i, z_i) three of the latter equations read

$$\dot{\boldsymbol{p}}_i + \boldsymbol{\omega}_{i0} \times \boldsymbol{p}_i = \boldsymbol{F}_i - \sum_{j=1}^{2}\begin{pmatrix} X_{ij} \\ Y_{ij} \\ Z_{ij}\end{pmatrix}, \tag{9-1}$$

$\boldsymbol{p}_i$ being the momentum of the wheelset with components related to this system and X_{ij}, Y_{ij}, Z_{ij} being the analogous contact force components; we take the signs of X_{ij}, Y_{ij}, Z_{ij} positive when they are opposite to o_ix_i, o_iy_i, o_iz_i. The angular velocity $\boldsymbol{\omega}_{i0}$ is given by (5-7).

From (5-1) we find for the components of the momentum

$$\boldsymbol{p}_i = m_i \boldsymbol{V}_i : \tag{9-2}$$

$$p_{xi} = m_i \dot{s}_i \{1 - r_0^{-1}(v_i \cos\varphi_0 - w_i \sin\varphi_0)\}, \quad p_{yi} = m_i \dot{v}_i, \quad p_{zi} = m_i \dot{w}_i. \tag{9-3}$$

For the forces X_{ij}, Y_{ij}, Z_{ij} we have

$$\begin{pmatrix} X_{ij} \\ Y_{ij} \\ Z_{ij} \end{pmatrix} = \begin{pmatrix} \cos(n_{ij}, x_i) & \cos(t_{xij}, x_i) & \cos(t_{yij}, x_i) \\ \cos(n_{ij}, y_i) & \cos(t_{xij}, y_i) & \cos(t_{yij}, y_i) \\ \cos(n_{ij}, z_i) & \cos(t_{xij}, z_i) & \cos(t_{yij}, z_i) \end{pmatrix} \begin{pmatrix} N_{ij} \\ T_{xij} \\ T_{yij} \end{pmatrix}, \tag{9-4}$$

so that because of (3-16):

$$\left.\begin{aligned} X_{ij} &= T_{xij} + r_0^{-1} x_{ij}(\mp N_{ij} \sin\beta_{ij} - T_{yij} \cos\beta_{ij}), \\ Y_{ij} &= \pm N_{ij} \sin\gamma_{ij} + T_{yij} \cos\gamma_{ij} + r_0^{-1} x_{ij} T_{xij} \cos\varphi_0, \\ Z_{ij} &= N_{ij} \cos\gamma_{ij} \mp T_{yij} \sin\gamma_{ij} - r_0^{-1} x_{ij} T_{xij} \sin\varphi_0. \end{aligned}\right\} \tag{9-5}$$

Altogether the equations of motion (1) become

$$m_i \ddot{s}_i \{1 - r_0^{-1}(v_i \cos\varphi_0 - w_i \sin\varphi_0)\} - 2 m_i \dot{s}_i r_0^{-1}(\dot{v}_i \cos\varphi_0 - \dot{w}_i \sin\varphi_0)$$
$$= F_{xi} - \sum_{j=1}^{2} T_{xij} - r_0^{-1} \sum_{j=1}^{2} (-1)^j N_{ij} x_{ij} \sin\beta_{ij} + r_0^{-1} \sum_{j=1}^{2} T_{yij} x_{ij} \cos\beta_{ij}, \tag{9-6a}$$

$$m_i \ddot{v}_i + m_i \dot{s}_i^2 r_0^{-1} \cos\varphi_0 = F_{yi} + \sum_{j=1}^{2} (-1)^j N_{ij} \sin\gamma_{ij} - \sum_{j=1}^{2} T_{yij} \cos\gamma_{ij} - r_0^{-1} \sum_{j=1}^{2} T_{xij} x_{ij} \cos\varphi_0, \tag{9-6b}$$

$$m_i \ddot{w}_i - m_i \dot{s}_i^2 r_0^{-1} \sin\varphi_0 = F_{zi} - \sum_{j=1}^{2} N_{ij} \cos\gamma_{ij} - \sum_{j=1}^{2} (-1)^j T_{yij} \sin\gamma_{ij} + r_0^{-1} \sum_{j=1}^{2} T_{xij} x_{ij} \sin\varphi_0. \tag{9-6c}$$

The three other equations of motion are the equations about the rate of change of the moment of momentum. They best can be written down for the moving coordinate system $(o_i^*, x_i^*, y_i^*, z_i^*)$:

$$\dot{\boldsymbol{b}}_i^* + \boldsymbol{\omega}_{i0}^* \times \boldsymbol{b}_i^* = \tilde{G}_i \boldsymbol{M}_i - \sum_{j=1}^{2} \boldsymbol{r}_{ij}^* \times \begin{pmatrix} X_{ij}^* \\ Y_{ij}^* \\ Z_{ij}^* \end{pmatrix}. \tag{9-7}$$

Here the moment of momentum is given by

$$\tilde{\boldsymbol{b}}_i^* = (J_i \omega_{xi}^*, J_{yi} \omega_{yi}^*, J_i \omega_{zi}^*), \tag{9-8}$$

so that we have, because of (5-3):

$$\left.\begin{aligned} b_{xi}^* &= J_i(r_0^{-1} \dot{s}_i \sin\varphi_0 \sin\psi_i + \dot{\varphi}_i), \\ b_{yi}^* &= J_{yi}\{r_0^{-1} \dot{s}_i(\sin\varphi_0 \cos\varphi_i \cos\psi_i + \cos\varphi_0 \sin\varphi_i) - \dot{\theta}_i + \dot{\psi}_i \sin\varphi_i\}, \\ b_{zi}^* &= J_i\{r_0^{-1} \dot{s}_i(-\sin\varphi_0 \sin\varphi_i \cos\psi_i + \cos\varphi_0 \cos\varphi_i) + \dot{\psi}_i \cos\varphi_i\}. \end{aligned}\right\} \tag{9-9}$$

The angular velocity $\boldsymbol{\omega}_{i0}^*$ of the system $(o_i^*, x_i^*, y_i^*, z_i^*)$ is given by (5-5). The vector $\boldsymbol{r}_{ij}^*$ of the contact point is determined by (4-17) and X_{ij}^*, Y_{ij}^*, Z_{ij}^* are the components of the contact point forces in the directions of the system $(o_i^*, x_i^*, y_i^*, z_i^*)$, so that from (3-32)

$$X_{ij}^* = \pm N_{ij} \sin\gamma_{ij} \sin\psi_i + T_{xij} \cos\psi_i + T_{yij} \cos\gamma_{ij} \sin\psi_i +$$
$$+ r_0^{-1} x_{ij}(\mp N_{ij} \sin\beta_{ij} \cos\psi_i + T_{xij} \cos\varphi_0 \sin\psi_i - T_{yij} \cos\beta_{ij} \cos\psi_i), \tag{9-10a}$$

$$Y_{ij}^* = N_{ij}(\cos\gamma_{ij} \sin\varphi_i \pm \sin\gamma_{ij} \cos\varphi_i \cos\psi_i) - T_{xij} \cos\varphi_i \sin\psi_i + T_{yij}(\cos\gamma_{ij} \cos\varphi_i \cos\psi_i \mp \sin\gamma_{ij} \sin\varphi_i) +$$
$$+ r_0^{-1} x_{ij}\{\pm N_{ij} \sin\beta_{ij} \cos\varphi_i \sin\psi_i + T_{xij}(\cos\varphi_0 \cos\varphi_i \cos\psi_i - \sin\varphi_0 \sin\varphi_i) + T_{yij} \cos\beta_{ij} \cos\varphi_i \sin\psi_i\}, \tag{9-10b}$$

$$Z_{ij}^* = N_{ij}(\cos\gamma_{ij} \cos\varphi_i \mp \sin\gamma_{ij} \sin\varphi_i \cos\psi_i) + T_{xij} \sin\varphi_i \sin\psi_i - T_{yij}(\cos\gamma_{ij} \sin\varphi_i \cos\psi_i \pm \sin\gamma_{ij} \cos\varphi_i) +$$
$$+ r_0^{-1} x_{ij}\{\mp N_{ij} \sin\beta_{ij} \sin\varphi_i \sin\psi_i - T_{xij}(\cos\varphi_0 \sin\varphi_i \cos\psi_i + \sin\varphi_0 \cos\varphi_i) - T_{yij} \cos\beta_{ij} \sin\varphi_i \sin\psi_i\}. \tag{9-10c}$$

By means of (5-5) and (8)—(10c) we can work up the equations (7) into

$$J_i \ddot{\varphi}_i + (J_i - J_{yi}) \dot{\psi}_i^2 \cos\varphi_i \sin\varphi_i + J_{yi} \dot{\theta}_i \dot{\psi}_i \cos\varphi_i + J_i r_0^{-1}(\ddot{s}_i \sin\psi_i + \dot{s}_i \dot{\psi}_i \cos\psi_i) \sin\varphi_0 +$$
$$+ J_i r_0^{-1} \dot{s}_i \dot{\psi}_i (\sin\varphi_0 \cos 2\varphi_i \cos\psi_i + \cos\varphi_0 \sin 2\varphi_i) + J_{yi} r_0^{-1} \dot{s}_i \dot{\theta}_i(\cos\varphi_0 \cos\varphi_i - \sin\varphi_0 \sin\varphi_i \cos\psi_i) -$$
$$- J_{yi} r_0^{-1} \dot{s}_i \dot{\psi}_i(\sin\varphi_0 \cos 2\varphi_i \cos\psi_i + \cos\varphi_0 \sin 2\varphi_i) =$$
$$= M_{xi} \cos\psi_i + M_{yi} \sin\psi_i - \sum_{j=1}^{2} N_{ij}\{(-1)^j b_{yzij} \sin\gamma_{ij} \cos\psi_i + a_{yzij} \cos\gamma_{ij}\} - \sum_{j=1}^{2} T_{xij} b_{yzij} \sin\psi_i +$$
$$+ \sum_{j=1}^{2} T_{yij} \{b_{yzij} \cos\gamma_{ij} \cos\psi_i - (-1)^j a_{yzij} \sin\gamma_{ij}\} - r_0^{-1} \sum_{j=1}^{2} (-1)^j N_{ij} x_{ij} b_{yzij} - \sin\psi_i \sin\beta_{ij} +$$
$$+ r_0^{-1} \sum_{j=1}^{2} T_{xij} x_{ij}(b_{yzij} \cos\varphi_0 \cos\psi_i + a_{yzij} \sin\varphi_0) + r_0^{-1} \sum_{j=1}^{2} T_{yij} x_{ij} b_{yzij} \sin\psi_i \cos\beta_{ij}, \tag{9-11a}$$

$$J_{yi}(-\ddot{\theta}_i + \ddot{\psi}_i \sin\varphi_i + \dot{\varphi}_i \dot{\psi}_i \cos\varphi_i) + J_{yi} r_0^{-1} \ddot{s}_i(\sin\varphi_0 \cos\varphi_i \cos\psi_i + \cos\varphi_0 \sin\varphi_i) +$$
$$+ J_{yi} r_0^{-1} \dot{s}_i \dot{\varphi}_i(\cos\varphi_0 \cos\varphi_i - \sin\varphi_0 \sin\varphi_i \cos\psi_i) - J_{yi} r_0^{-1} \dot{s}_i \dot{\psi}_i \sin\varphi_0 \cos\varphi_i \sin\psi_i =$$
$$= (-M_{xi} \sin\psi_i + M_{yi} \cos\psi_i) \cos\psi_i + M_{zi} \sin\varphi_i - \sum_{j=1}^{2} T_{xij} b_{xzij}$$
$$- \sum_{j=1}^{2} T_{yij}\{a_{xzij} \cos\gamma_{ij} - (-1)^j x_{ij}^* \sin\gamma_{ij} \cos\varphi_i\} - r_0^{-1} \sum_{j=1}^{2} T_{xij} x_{ij}(a_{xzij} \cos\varphi_0 + x_{ij}^* \sin\varphi_0 \cos\varphi_i) +$$
$$+ r_0^{-1} \sum_{j=1}^{2} T_{yij} x_{ij} b_{xzij} \cos\beta_{ij} + \left[\sum_{j=1}^{2} N_{ij}\{(-1)^j a_{xzij} \sin\gamma_{ij} + x_{ij}^* \cos\gamma_{ij} \cos\varphi_i\} - r_0^{-1} \sum_{j=1}^{2} (-1)^j N_{ij} x_{ij} b_{xzij} \sin\beta_{ij}\right], \tag{9-11b}$$

$$\begin{aligned} & J_i\ddot{\psi}_i \cos\varphi_i - 2J_i\dot{\varphi}_i\dot{\psi}_i \sin\varphi_i - J_{yi}\dot{\varphi}_i\dot{\theta}_i + J_{yi}\dot{\varphi}_i\dot{\psi}_i \sin\varphi_i + \\ & \quad + J_i r_0^{-1}\ddot{s}_i(\cos\varphi_0 \cos\varphi_i - \sin\varphi_0 \sin\varphi_i \cos\psi_i) - 2J_i r_0^{-1}\dot{s}_i\dot{\varphi}_i(\sin\varphi_0 \cos\varphi_i \cos\psi_i + \cos\varphi_0 \sin\varphi_i) + \\ & \quad + J_{yi} r_0^{-1}\dot{s}_i(-\dot{\theta}_i + \dot{\psi}_i \sin\varphi_i) \sin\varphi_0 \sin\psi_i + J_{yi} r_0^{-1}\dot{s}_i\dot{\varphi}_i(\cos\varphi_0 \sin\varphi_i + \sin\varphi_0 \cos\varphi_i \cos\psi_i) = \\ & = (M_{xi}\sin\psi_i - M_{yi}\cos\psi_i)\sin\varphi_i + M_{zi}\cos\varphi_i + \sum_{j=1}^{2} (-1)^j N_{ij}(a_{xyij}\sin\gamma_{ij} - x_{ij}^*\cos\gamma_{ij}\sin\varphi_i) + \\ & \quad + \sum_{j=1}^{2} T_{xij}b_{xyij} - \sum_{j=1}^{2} T_{yij}\{a_{xyij}\cos\gamma_{ij} + (-1)^j x_{ij}^* \sin\gamma_{ij}\sin\varphi_i\} + r_0^{-1}\sum_{j=1}^{2}(-1)^j N_{ij}x_{ij}b_{xyij}\sin\beta_{ij} - \\ & \quad - r_0^{-1}\sum_{j=1}^{2} T_{xij}(a_{xyij}\cos\varphi_0 - x_{ij}^*\sin\varphi_0\sin\varphi_i) - r_0^{-1}\sum_{j=1}^{2} T_{yij}b_{xyij}\cos\beta_{ij}\,. \end{aligned} \tag{9-11c}$$

It is possible to subject the equations of motion (6a—c) and (11a—c) to a linear transformation in such a way that they become identical to the Lagrangian equations (8-4a)—(8-5c). One easily finds that, symbolically, the transformation can be expressed by the relation

$$\begin{Bmatrix} (8\text{-}4\,a) \\ (8\text{-}4\,b) \\ (8\text{-}4\,c) \\ (8\text{-}5\,a) \\ (8\text{-}5\,b) \\ (8\text{-}5\,c) \end{Bmatrix} = \boldsymbol{M} \begin{Bmatrix} (9\text{-}6\,a) \\ (9\text{-}6\,b) \\ (9\text{-}6\,c) \\ (9\text{-}11\,a) \\ (9\text{-}11\,b) \\ (9\text{-}11\,c) \end{Bmatrix}, \tag{9-12}$$

where

$$\boldsymbol{M} = \begin{Bmatrix} m_{11} & 0 & 0 & m_{14} & m_{15} & m_{16} \\ 0 & 1 & 0 & 0 & 0 & 0 \\ 0 & 0 & 1 & 0 & 0 & 0 \\ 0 & 0 & 0 & 1 & 0 & 0 \\ 0 & 0 & 0 & 0 & -1 & 0 \\ 0 & 0 & 0 & 0 & \sin\varphi_i & \cos\varphi_i \end{Bmatrix}, \tag{9-13}$$

$$\left.\begin{aligned} m_{11} &= 1 - r_0^{-1}(v_i\cos\varphi_0 - w_i\sin\varphi_0)\,, \\ m_{14} &= r_0^{-1}\sin\varphi_0\sin\psi_i\,, \\ m_{15} &= r_0^{-1}(\sin\varphi_0\cos\varphi_i\cos\psi_i + \cos\varphi_0\sin\varphi_i)\,, \\ m_{16} &= r_0^{-1}(-\sin\varphi_0\sin\varphi_i\cos\psi_i + \cos\varphi_0\cos\varphi_i)\,. \end{aligned}\right\} \tag{9-14}$$

In this way a second check on the equations of motion is obtained.

10. The motion of the wheelset; concluding remarks

The motion of the wheelset is determined by the six equations (8-4a)—(8-5b) or the six equations (9-6 a—c) and (9-11 a—c), as the case may be. However, the unknowns are not only the six coordinates s_i, v_i, w_i, φ_i, θ_i, ψ_i but also the two normal forces N_{ij} and the four tangential forces T_{xij}, T_{yij}: althogether twelve unknowns.

But besides the equations of motion we dispose on the two constraint relations (4-21) and on four relations for the tangential forces, which are completely determined, once the normal forces N_{ij}, the principal radii of curvature of the rail and the wheel in the contact point and also the creepage and spin quantities

$$v_{xij} = W_{txij}/\dot{s}_i\,, \qquad v_{yij} = W_{tyij}/\dot{s}_i\,, \qquad \varphi_{ij} = \omega_{nij}/\dot{s}_i \tag{10-1}$$

are known [7]. For vanishing creepage and spin the expressions for the tangential forces are given in [5 p. 404—405].

In many cases the equations of motion can be considerably reduced by taking into account that

$$\left.\begin{aligned} & |v_i| \ll b\,, \quad |w_i| \ll b\,, \quad |\varphi_i| \ll 1\,, \quad |\psi_i| \ll 1\,, \\ & |x_{ij}| \ll b\,, \quad |b \mp y_{ij}| \ll b\,, \quad |r - z_{ij}| \ll b\,, \\ & |x_{ij}^*| \ll b\,, \quad |b \mp y_{ij}^*| \ll b\,, \quad |r - z_{ij}^*| \ll b\,, \\ & b \ll r_0\,. \end{aligned}\right\} \tag{10-2}$$

Along these lines it is possible to develop a so-called first order theory, as it has been done already for the motion on a tangent track [5 p. 403—405]. The simplified equations of motion and the constraint relations remain essentially non-linear because of the conicity angles γ_{ij}, γ_{ij}^*, which still can have any given values. Moreover, the relations for the tangential forces mostly have to be considered as non-linear because of the magnitude of the creep and spin quantities.

An important special case is that in which the motion through the curve is stationary, so that

$$\ddot{s}_i = 0\,, \qquad \dot{v}_i = 0\,, \qquad \dot{w}_i = 0\,, \qquad \dot{\varphi}_i = 0\,, \qquad \dot{\theta}_i = 0\,, \qquad \dot{\psi}_i = 0\,. \tag{10-3}$$

In that case it is possible to determine $\dot{s}_i$, v_i, w_i, φ_i, $\dot{\theta}_i$ and ψ_i for any combination of F_{xi}, F_{yi}, F_{zi}, M_{xi}, M_{yi}, M_{zi}. Moreover, the general equations enable us to discuss the stability of such a stationary motion.

The first-order theory and the stationary motion will be the subject of a forthcoming publication.

References

1 HEUMANN, H., Grundzüge der Führung der Schienenfahrzeuge, München (no year indicated), 156 S. (reprint from „Elektrische Bahnen“ 1950—1953).
2 BORGEAUD, G., Le passage en courbes de véhicules de chemin de fer, dont les essieux fournissent un effort de traction continu, Thèse Zürich (1937), 169 S.
3 MÜLLER, C. TH., Der Eisenbahnradsatz/Kinematik, Spurführungsgeometrie und Führungsvermögen, Glasers Annalen 77 (1953) 9, S. 264—281, 10, S. 315.
4 ALDINGTON, T. A., private communication.
5 PATER, A. D. DE, The exact theory of the motion of a single wheelset moving on a perfectly straight track, Proc. 6th IAVSD-Symposium held at Berlin on September 3—7, 1979, Lisse (1980) S. 397—407.
6 PATER, A. D. DE, The exact theory of the motion of a single wheelset moving on a purely straight track, Delft University of Technology, Dept. of Mech. Engg., Lab. of Engg. Mechs. Report No. 648 (1979), 5 + 78 S.
7 KALKER, J. J., Survey of Wheel-Rail Rolling Contact Theory, Vehicle Systems Dynamics 8 (1979) 4, S. 317—358.

Eingereicht am 12. 5. 1980

Anschrift: Prof. dr. ir. A. D. DE PATER, Delft University of Technology, Laboratory of Engineering Mechanics of the Mechanical Engineering Department, 2628 CD Delft, Mekelweg 2, Niederlande

ZAMM 61, 293 – 298 (1981)

J. van der Zanden

On the Incompleteness of the Number of Physical Parameters of a Free Streamline Potential Flow Problem

The two-dimensional free-streamline solution of a plate rectangular in the flow of the jet from a nozzle is solved numerically. Earlier theoretical discussions of this problem mentioned only three physical parameters. However, this number must be equal to that of the independent geometrical parameters of the problem, which is four, otherwise the problem is not well posed. So a fourth parameter has been introduced. Numerical integration is preferred, although analytical integration is, in principle, possible. The determination of parameters and streamlines depends on the solution of nonlinear equations with inequality constraints.

Das zweidimensionale Umströmungsproblem einer ebenen Platte, die senkrecht von einem Strahl angeströmt wird, wird numerisch gelöst (freie Stromlinien-Lösung). Frühere theoretische Diskussionen dieses Problems berücksichtigten nur drei physikalische Parameter. Die Anzahl dieser Parameter muß jedoch wie die Anzahl der unabhängigen geometrischen Parameter vier sein, andernfalls ist die Aufgabe nicht korrekt gestellt. Daher wurde ein vierter Parameter eingeführt. Eine numerische Integration wird vorgezogen, obwohl prinzipiell eine analytische Lösung möglich wäre. Die Bestimmung der Parameter und der Stromlinien hängt von der Lösung eines nichtlinearen Gleichungssystems mit Ungleichungen als Nebenbedingungen ab.

Рассматривается численное решение двумерной задачи обтекания вертикальной пластины в потоке, которий выходит из канала в виде свободной струи. В литературе существуют теоретические решения задачи с тремя физическими параметрами, которые следует определить. Дано новое решение с дополнительным физическим параметром, которое наудено численно несмотря на то, что существует возможность аналитического решение. Определение параметров и линий тока ведет к решению нелинейных уравнений с дополнительными условиями в виде неравенств.

1. Introduction

The two-dimensional, incompressible, irrotational, non-viscous flow of a jet from a nozzle against a plate (see figure 1) has first been studied by Greenhill [1] who gives an analytical solution without going into the details of the determination of the parameters. In his solution only three physical parameters occur: the upstream velocity V (which is the minimum velocity along the walls of the nozzle), and the jet deflection angles α_1, α_2. Much later Birkhoff and Zarantonello [2] discuss this problem again and solve it in a different way. Again, only three unknown physical parameters occur in the solution (Chapter V, p. 110, equation (21)), and the length of the plate is the only geometrical quantity that has been expressed in terms of the solution. Complete calculations have not been performed and it might be due to this, that the fact that a complete description of the geometry requires four independent quantities has been overlooked.

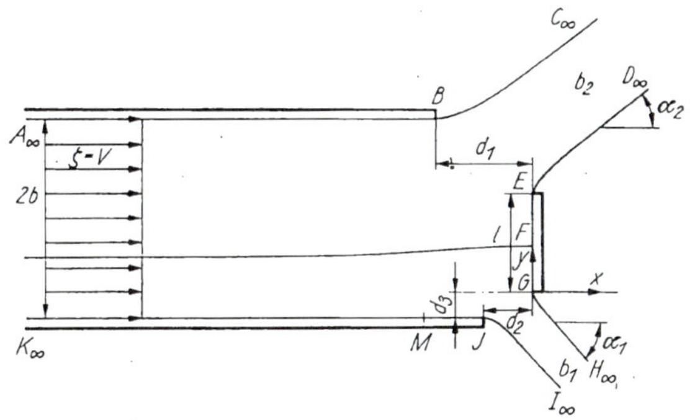

Figure 1. z-plane, $z = x + iy$

2. In search of a fourth physical parameter

The width of the nozzle (see figure 1) has been chosen as normalization factor, whence four independent geometrical quantities characterize the problem: l, d_1, d_2, d_3. They can be calculated from the coordinates z_B, z_E, z_G and z_J where the velocities are known: equal to one and parallel to the plate just left behind. With only three parameters not all possible sets of values for $\{l, d_1, d_2, d_3\}$ can be obtained. How to find a fourth parameter? Examine the case of a plate close to a wall (figure 2) in a semi-infinite stream, which is a limiting case of figure 1 ($b \to \infty$, $d_2 \to \infty$) with only two parameters l and d_3. The velocity V is no parameter any more and $\alpha_1 = \alpha_2 = 0$. The form of the dividing streamline (an S-curve) indicates a deceleration of the flow along the wall until a minimum V_M has been reached at M where upon the velocity increases up to V at I. The physical parameters are thus b_1, V_M. It is plausible to assume that the dividing streamline of the general problem has the same form and that along one of the walls of the nozzle the velocity is decelerated until some minimum V_M before it is accelerated to the value 1 at the end of the wall. Now the problem is well posed and the numerical calculations can be started.

Later on, the invention of V_M appeared to have been made long before. Gurevich [3] treats the problem of figure 2 for $\gamma = \pi/2$ and refers to Bonder [4] for the general case. Bonder solves this problem along the same lines as in our paper and refers to Villat [5] for another approach started in [6], [7], which is also capable of the treatment of curved obstacles. Discussing the case when the object in the flow is a wedge, Villat [6] remarks (p. 238) that the position of the fixed wall with respect to the wedge cannot be prescribed arbitrarily. He pays no more

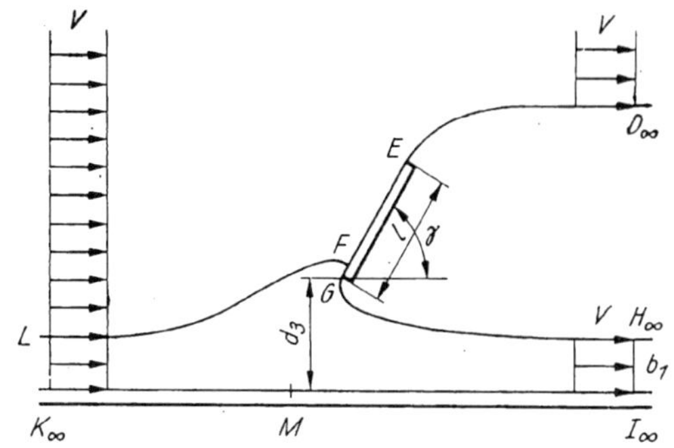

Figure 2. Plate and wall

attention to it. For the plate this anomaly has been removed by BONDER, who thus introduced V_M as a physical parameter of the problem. Independently, so it seems, GREEN [8] studied another case: an infinite plate gliding on a finite stream ($d_1 = -\infty$, $d_2 = \infty$, $l = \infty$).

GREEN [9] also studied the problem of a finite plate gliding on a finite stream ($d_1 = -\infty$, $d_2 = \infty$), again with the remark that along the bottom the flow has no constant velocity but goes through a minimum. Curiously enough BIRKHOFF and ZARANTONELLO (p. 108) refer to GREEN and BONDER for limiting cases (figure 2), and (p. 123) to Villat without mentioning the fundamental difference (the invention of V_M) of their solution.

For $\gamma = \pi/2$ the general problem can be solved without the use of elliptic functions. A complete numerical solution is given.

3. A plate in the flow of the jet from a nozzle

When a velocity potential $\varphi(x, y)$ and a streamfunction $\psi(x, y)$ are introduced, it is well known that the complex potential $\chi(z) = \varphi(x, y) + i\psi(x, y)$ is an analytic function of $z = x + iy$, so

$$\frac{d\chi}{dz} = v_x - iv_y = \zeta$$

with velocities v_x and v_y in x- and y-direction. Mapping the χ-plane onto the z-plane (or the reverse) yields the solution. When free streamlines are assumed, the geometry in the z-plane is unknown. However the ζ-plane (or Ω-plane with $\Omega = -\log \zeta$) can be mapped onto the same plane (an upper half λ-plane) as the flow region in the χ-plane, whence the solution is given by

$$z - z_0 = \int_{\lambda_0}^{\lambda} \frac{1}{\zeta(\lambda)} \frac{d\chi}{d\lambda} d\lambda . \tag{1}$$

The χ-plane is a notched infinite strip (see figure 3), the s-plane ($s = i\zeta$) is a notched semi circle (see figure 4), which can be mapped onto the notched upper-half T-plane (see figure 5) with JOUKOWSKI's transform $T = -(s + 1/s)/2$ as the notch remains a strait notch. It must be stressed that it is not known beforehand if M is on the wall AB or along KJ! With the SCHWARZ-CHRISTOFFEL transformation the polygonal flow regions of the χ-plane as well as the T-plane can be mapped onto an upper half λ-plane with a free choice of three points on the real λ-axis for each.

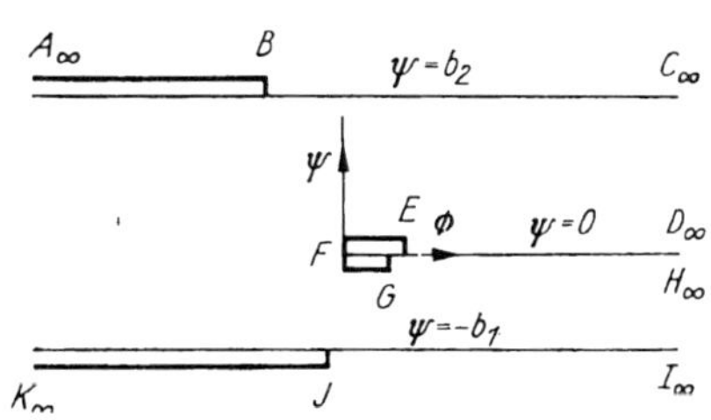

Fig. 3. χ-plane

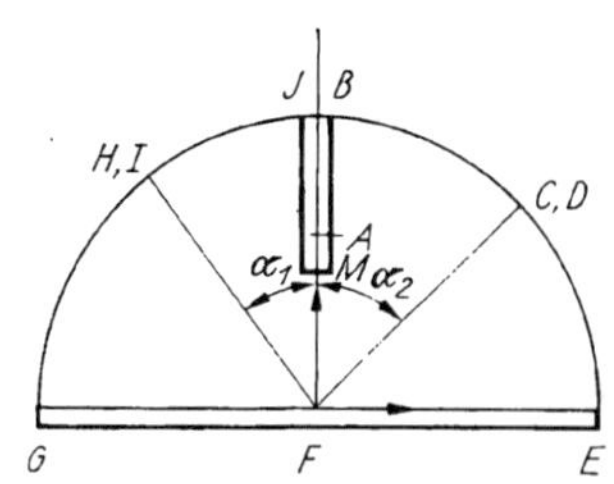

Fig. 4. s-plane

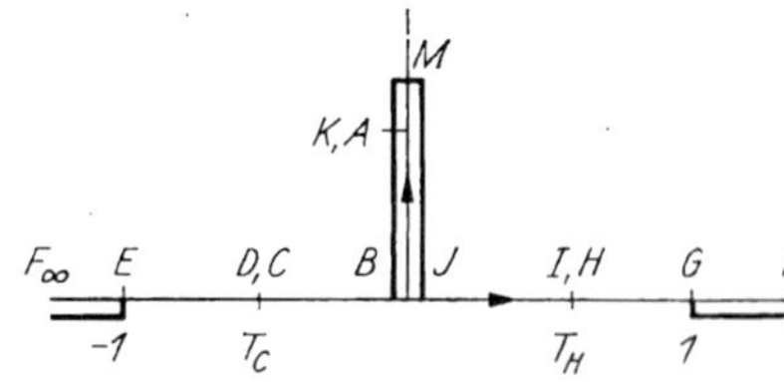

Fig. 5. T-plane

Choosing

$$\lambda_M = 0 , \quad \lambda_B = -d , \quad \lambda_J = d$$

the mapping of T onto λ can be performed with

$$\frac{dT}{d\lambda} = K_1 \frac{\lambda}{(\lambda + d)^{1/2} (\lambda - d)^{1/2}} = \pm K_1 \frac{\lambda}{(\lambda^2 - d^2)^{1/2}} \tag{2}$$

which can easily be integrated

$$T - T_0 = K_1(\lambda + d)^{1/2} (\lambda - d)^{1/2} + L_1 = \pm K_1(\lambda^2 - d^2)^{1/2} + L_1 . \tag{3}$$

In (2) as well as in (3) the minus sign has to be used in the second part of the expressions for values of λ that are both real and smaller than $-d$! In spite of this disadvantage that expression is used for computational purposes as the square root is to be evaluated only once. Application of the conditions on ζ, s, T and λ can be summarized as follows

$$s = i\zeta\,, \qquad s_M = iV_M\,, \qquad T = -\frac{1}{2}\left(s + \frac{1}{s}\right), \qquad T_M = ia_M = \frac{i}{2}\left(\frac{1}{V_M} - V_M\right),$$

$$T_E = -1\,, \qquad T_C = -\sin\alpha_1\,, \qquad T_A = ia = \frac{i}{2}\left(\frac{1}{V} - V\right), \qquad T_H = \sin\alpha_2\,, \qquad T_G = 1\,,$$

$$\left.\begin{array}{lll} T_B = 0\,, & \lambda_B = -d \Rightarrow L_1 = -T_0\,, & \\ T_G = 1\,, & \lambda_G = 1 \Rightarrow K_1 = (1-d^2)^{-1/2} & \\ T_M = ia_M\,, & \lambda_M = 0 \Rightarrow K_1 = a_M/d & \end{array}\right\} \Rightarrow d = a_M(1 + a_M^2)^{-1/2}.$$

Inversion of equation (3) leads to

$$\lambda = \pm\frac{1}{K_1}(T^2 + a_M^2)^{1/2} = \frac{1}{K_1}(T + ia_M)^{1/2}\,(T - ia_M)^{1/2}\,. \tag{4}$$

The minus sign has to be used for values of T that are both real and less than zero, but also for purely imaginary values of T, $0 < \operatorname{Im} T < ia_M$, which are between M and B. Therefore it can not be decided if λ_A is positive or negative:

$$\lambda_C = -\frac{1}{K_1}(\sin^2\alpha_2 + a_M^2)^{1/2}\,, \qquad \lambda_A = \pm\frac{1}{K_1}(a^2 + a_M^2)\,, \qquad \lambda_{\hat{H}} = \frac{1}{K_1}(\sin\alpha_1^2 + a_M^2)^{1/2}\,.$$

That is why it is wise to choose λ_A as unknown in place of the physical variable V as will be discussed later on.

The inverse of the JOUKOWSKI transform

$$s = -T \pm \sqrt{T^2 - 1} = -T + (T+1)^{1/2}\,(T-1)^{1/2}\,,$$
$$1/s = -T \mp \sqrt{T^2 - 1} = -T - (T+1)^{1/2}\,(T-1)^{1/2}, \tag{5}$$

also shows the choice of sign for T both real and less than -1 if only one evaluation of a square root is chosen.

For the mapping of the flow region in the χ-plane onto the upper half λ-plane the vertices χ_C, χ_F, and χ_H are chosen to correspond with λ_C, ∞, λ_H as already found from the $T - \lambda$ mapping. The vertex χ_A is mapped onto λ_A' which is not necessarily equal to λ_A.

$$\frac{d\chi}{d\lambda} = \frac{K_2}{(\lambda - \lambda_C)\,(\lambda - \lambda_A')\,(\lambda - \lambda_H)} = K_2\left(\frac{c}{\lambda - \lambda_C} + \frac{a}{\lambda - \lambda_A'} + \frac{h}{\lambda - \lambda_H}\right), \tag{6}$$

with

$$a + c + h = 0\,, \qquad \text{etc.,} \qquad a = \frac{1}{(\lambda_A' - \lambda_C)\,(\lambda_A' - \lambda_H)}\,, \qquad \text{etc.}\,,$$

which can easily be integrated

$$\chi - \chi_0 = K_2(c\log(\lambda - \lambda_C) + a\log(\lambda - \lambda_A') + h\log(\lambda - \lambda_H)) + L_2\,, \tag{7}$$

when for the multiple valued logarithm an argument in $(-\pi, \pi]$ is chosen. The conditions on χ lead to

$$\lambda \to \lambda_F = \infty\,, \qquad \chi_F = 0\,, \qquad a + c + h = 0 \Rightarrow L_2 = -\chi_0\,,$$
$$\lambda \text{ real}\,, \qquad \lambda_C < \lambda < \lambda_A'\,, \qquad \psi = b_2 \Rightarrow K_2\pi(a + h) = b_2\,,$$
$$\lambda \text{ real}\,, \qquad \lambda_A' < \lambda < \lambda_H\,, \qquad \psi = -b_1 \Rightarrow K_2\pi h = -b_1\,,$$
$$\text{continuity equation} \quad (\Delta\varphi = 0) \Rightarrow b_1 + b_2 = 2bV\,, \qquad \text{consistency} \quad \lambda \to \lambda_A\,, \quad \varphi \to -\infty \Rightarrow \lambda_A' = \lambda_A\,,$$

leading to

$$K_2 = 2bV/\pi a\,, \qquad b_1 = 2bVh/a\,, \qquad b_2 = 2bVc/a = 2bV - b_1\,,$$

which shows that b_1 and b_2 depend on the physical variables α_1, α_2, V_M, V.

The solution is complete as can be seen from inspection of (1), (3), (5) and (6). The integral (1) can be evaluated in the s-plane, T-plane or λ-plane. In terms of λ analytical integration depends on evaluation of integrals of the type

$$\int \frac{(\lambda^2 - p^2)^{1/2}}{\lambda - q}\,d\lambda\,,$$

for which formulae can be found in BRONSHTEIN-SEMENDYAYEV [10], p. 430—432. However, the answer is in terms of multiple valued functions and therefore requires accurate bookkeeping of the arguments of the complex quantities. This problem can be avoided when the integrals are evaluated numerically, which is most easy in the upper half λ-plane, where ζ and $d\chi/d\lambda$ are single valued (no notches!).

The ROMBERG algorithm (see RALSTON and WILF [11]) has been chosen with automatic steplength and accuracy control (see PRAGER [12]).

4. Determination of the physical parameters

When the width of the channel (2b) is used as characteristic quantity (in order to make others dimensionless), the geometry is completely determined by l, d_1, d_2, d_3 (see figure 1). If these are given, the physical parameters α_1, α_2, V_M, V can be determined from (1)

$$l'(\alpha_1, \alpha_2, V_M, V) = \mathrm{Im}\,(z_E - z_G)\,, \qquad l - l'(\alpha_1, \alpha_2, V_M, V) = 0\,, \quad \text{etc.}$$

It has been remarked in the preceding paragraph that the sign of λ_A cannot be decided, due to the fact that the position of M can be on the upper plate or on the lower plate of the nozzle.

This difficulty is avoided when λ_A is chosen as physical parameter instead of V, which is now a function of V_M and λ_A

$$V = -K_1(d^2 - \lambda_A^2)^{1/2} + K_1(1 - \lambda_A^2)^{1/2}\,,$$

as both K_1 and d are functions of $a_M = (1/V_M - V_M)/2$ only. There are four equations to be solved with inequality constraints

$$0 < \alpha_1 < \pi/2\,, \qquad 0 < \alpha_2 < \pi/2\,, \qquad 0 < V_M < 1\,, \qquad -d < \lambda_A < d\,. \tag{8}$$

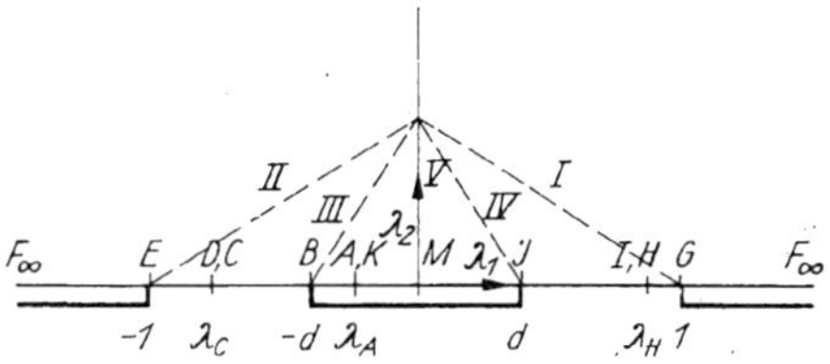

Fig. 6. Integration in λ-plane, $\lambda = \lambda_1 + i\lambda_2$

Paths of integration in the λ-plane must avoid the singular points λ_C, λ_A and λ_H. With $z_0 = z_G$, for $\lambda = 1$, the path of integration for $l'(\alpha_1, \alpha_2, V_M, \lambda_A)$ is I + II (see figure 6). For the quantities d_1', d_2', d_3', z_B and z_J have to be calculated by integration of (1) along I + III and I + IV. If the solution is found, the position of M can be calculated (I + V). The parameters α_1, α_2, V_M and λ_A are to be found from the nonlinear equations

$$l'(\alpha_1, \alpha_2, V_M, \lambda_A) - l = 0\,, \qquad d_i'(\alpha_1, \alpha_2, V_M, \lambda_A) - d = 0\,, \qquad i = 1, 2, 3\,, \tag{9}$$

with inequality constraints (8), or from minimization of

$$E = \left\{(l' - l)^2 + \sum_{i=1}^{3} (d_i' - d_i)^2\right\}^{1/2} \tag{10}$$

with the same inequality constraints. If a gradient algorithm to solve (9) is not convergent for a chosen initial estimate, this estimate can be improved by minimizing (10) and (8) with an ever convergent direct search algorithm. For references see VAN DER ZANDEN [13], [14]. Computer time can be saved by requiring an integration accuracy that is proportional with the accuracy of the equations (use (10)) attained. In our calculations 0.2 E was used. The solution was supposed to be found accurately enough for $E \leq 0.001$. For results see table 1. A control of the calculations is very important. A symmetric case, no 1 in table 1, should give exactly the same answers as the example of RÉTHY flow (see [13], [14]), which proved to be true. Cases which can be transformed onto another by symmetry operations, like 2 and 3 of table 1, can also be used for control:

$$\alpha_1^{(2)} = \alpha_2^{(3)}\,, \qquad \alpha_2^{(2)} = \alpha_1^{(3)}\,, \qquad V_M^{(2)} = V_M^{(3)}\,, \qquad \lambda_A^{(2)} = -\lambda_A^{(3)}\,, \quad \text{etc.}$$

Cases with large negative values for d_2 and d_3 ($\simeq -2l$) must be comparable with the flow of a free jet against a plate (see preceding paper) with $V_M \to 1$, $\lambda_A \to 0$ which proved to be true (case no. 5). Convergence of integration is troublesome for α_1, α_2 close to 0 or $\pi/2$. In such cases an infinite wall or a semi-infinite plate can be chosen with one physical parameter less, i.e. construct a solution for such special cases (no. 7, 11, 15). For a good initial estimate, $E = O(0.5)$, about 7000–10000 evaluations of the integrand of (1) are necessary in order to make $E \leq 0.0001$.

On the IBM 370/158 of the Delft University of Technology this is equivalent with about 7–10 seconds. If $E = O(3)$ some more time is necessary ranging from 5 to 20 seconds.

Equipotential lines and streamlines can now be calculated. The procedure is described in the preceding paper. A computing time of about 15 seconds for the calculation of about 250 coordinates z sufficed. Both λ-plane and z-plane are given in figures 7 and 8.

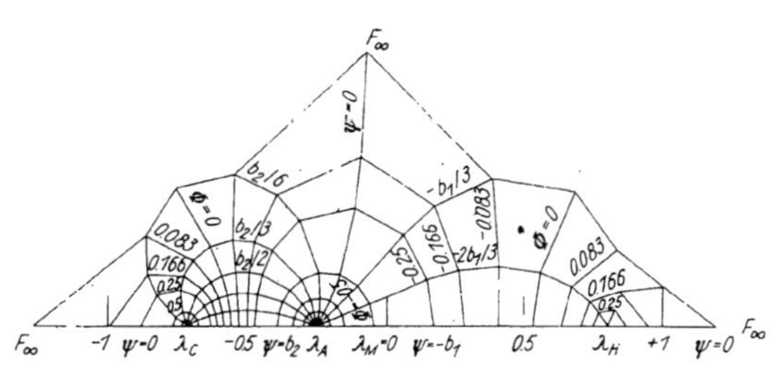

Fig. 7. Streamlines and equipotential lines in λ-plane $l = 1$, $d_1 = -1$, $d_2 = -0.5$, $d_3 = -0.25$ (no. 12 of table 1)

Table 1. Parameters. Plate in flow of jet from nozzle

No.	l	d_1	d_2	d_3	α_1	α_2	λ_A	V_M	V	b_1	Im (z_F)	Re (z_M)	Im (z_M)	E
1	2.0	−1.0	−1.0	0.0	1.11087	1.11088	0.00000	0.57796	0.57796	0.57796	1.00000	$-\infty$		$2.2 \cdot 10^{-5}$
2	2.0	−1.0	−1.0	−0.5	0.77703	1.39505	0.06308	0.62143	0.62425	0.74679	0.70382	−2.24149	1.50000	$2.8 \cdot 10^{-5}$
3	2.0	−1.0	−1.0	0.5	1.39506	0.77703	−0.06309	0.62141	0.62423	0.50168	1.29616	−2.24139	0.50000	$5.4 \cdot 10^{-5}$
4	2.0	−3.0	−1.0	0.5	1.39705	0.76774	−0.26882	0.73943	0.88543	0.47269	1.21195	−1.42240	0.50004	$9.4 \cdot 10^{-5}$
5	2.0	−4.0	−4.0	0.5	1.17361	0.73745	−0.00002	0.99250	0.99250	0.83704	1.23793	−7.77399	0.49999	$3.7 \cdot 10^{-4}$
6	3.0	−1.0	−1.0	0.0	1.11798	1.52364	0.02424	0.55188	0.55218	0.58618	1.06171	−2.96503	2.00000	$7.3 \cdot 10^{-5}$
7	3.0	−1.0	−1.0	−0.5	0.78443	1.55996	0.06088	0.62002	0.62262	0.74493	0.71625	−2.27548	1.50929	$2.0 \cdot 10^{-2}$
8	1.0	−1.0	−1.0	−0.25	0.86641	0.58729	−0.03899	0.72675	0.72855	0.59573	0.55352	−2.32351	−0.25005	$5.5 \cdot 10^{-5}$
9	1.0	−0.5	−1.0	−0.25	0.89055	0.48717	0.07258	0.64977	0.65404	0.61538	0.60099	−1.89004	1.74996	$3.9 \cdot 10^{-5}$
10	1.0	0.0	−1.0	−0.25	0.91266	0.28084	0.18578	0.57737	0.59860	0.63188	0.65207	−1.25282	1.74999	$3.2 \cdot 10^{-5}$
11	1.0	0.5	−1.0	−0.25	0.91997	0.09462	0.22676	0.55250	0.58164	0.63716	0.67095	−1.11033	1.74997	$5.8 \cdot 10^{-5}$
12	1.0	−1.0	−0.5	−0.25	0.82141	0.64098	−0.25425	0.58750	0.63193	0.38286	0.46834	−1.12125	−0.25001	$9.5 \cdot 10^{-6}$
13	1.0	−0.5	−0.5	−0.25	0.84089	0.52973	−0.14913	0.52798	0.53875	0.38746	0.51489	−1.44378	−0.25002	$1.7 \cdot 10^{-5}$
14	1.0	0.0	−0.5	−0.25	0.85855	0.30274	−0.03777	0.47090	0.47143	0.39122	0.56635	−2.30012	−0.24979	$3.8 \cdot 10^{-4}$
15	1.0	0.5	−0.5	−0.25	0.01000	0.09303	−0.33820	0.29098	0.31299	0.38286	0.46834	−1.12125	−0.25001	$2.4 \cdot 10^{-1}$

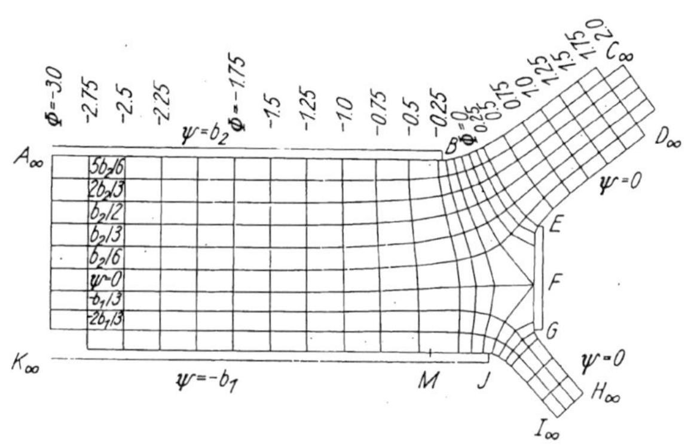

Fig. 8. Streamlines and equipotential lines in z-plane $l = 1$, $d_1 = -1$, $d_2 = -0.5$, $d_3 = -0.25$ (no. 12 of table) 1.

Acknowledgement

Thanks are due to H. BUITENHUIS and A. K. CHESTERS for clarifying discussions about the missing physical parameter.

References

1 GREENHILL, G., Report on the theory of a streamline past a plane barrier and the discontinuity arising at the edge with an application of the theory to an aeroplane, HMSO, Advis. Comm. Aeron. RM **19** (1910).
2 BIRKHOFF, G., ZARANTONELLO, E. H., Jets, wakes, and cavities, New York, Academic Press (1957).
3 GUREVICH, M. I., Theory of jets in ideal fluids, New York, Academic Press (first published in russian in 1961) (1965).
4 BONDER, J., Sur un cas de mouvements plans à deux sillages, Ann. Acad. Sci. Techn. Warsaw **3**, 155 —192 (1936).
5 VILLAT, H., Sur un calcul de résistance dans un courant fluide limité par un mur, Ann. Sci. de l'École Norm. Sup. **35**, 251—312 (1918).
6 VILLAT, H., Sur la résistence des fluides, Ann. Sci. de l'École Norm. Sup. **28**, 203—311 (1911).
7 VILLAT, H., Sur le mouvement d'un solide donnée dans un fluide limité par une paroie fixe, J. de Math. Pures et Appl. (Ser. 6) **7**, 353—408(1911).
8 GREEN, A. E., The gliding of a plate on a stream of finite depth, Proc. Cambr. Philos. Soc. **31**, 589—603 (1935). Corrigendum, Proc. Cambr. Philos. Soc. **32**, 183 (1936).
9 GREEN, A. E., The gliding of a plate on a stream of finite depth, Part II, Proc. Cambr. Philos. Soc. **32**, 67—85 (1936).
10 BRONSHTEIN-SEMENDYAYEV, A guide book to mathematics, Zürich (Deutsch)/New York, Springer (translated from russian) (1973).
11 RALSTON, A., WILF, H. S., (eds.), Mathematical methods for digital computers, Vol. 2, New York, Wiley (1967).
12 PRAGER, W., Introduction to basic FORTRAN programming and numerical methods, New York, Blaisdell (1965).
13 VAN DER ZANDEN, J., Numerical calculation of two classes of potential problems, illustrated with the flow from a channel against a wedge, Delft, University of Technology, Dept. Mech. Engng. WTHD 71 (1975).
14 VAN DER ZANDEN, J., On the numerical solution of some problems of free streamline potential flow theoiy, ZAMM, **59**, 748—752 (1979).

Eingereicht am 5. 11. 1979

Anschrift: Ing. J. P. P. M. VAN DER ZANDEN, Technische Hogeschool Delft, Department of Mechanical Engineering, Fluid Dynamics Group, Rotterdamseweg 145, 2628 AL Delft, Niederlande

ZAMM 61, 299–303 (1981)

R. C. GUPTA

Hydrodynamic Inlet Region Flow of Power-Law Fluids in a Circular Tube

Die Impuls-Energie-Integral-Methode zur Berechnung laminarer Rohreinlaufströmungen von Potenzgesetz-Flüssigkeiten wurde durch Verwendung der Wolkowschen Erweiterung der Karman-Pohlhausen-Methode verbessert. Es werden analytische Ausdrücke für die Grenzschichtentwicklung und Druckverteilung im Einlaufbereich angegeben. Die Resultate werden grafisch dargestellt und mit früheren Ergebnissen verglichen.

Momentum-energy-integral analysis for the laminar flow of power-law fluids in the inlet region of a circular tube has been improved by incorporating Volkov's refinement of the Karman-Pohlhausen method. Analytical expressions for the boundary layer development and the inlet region pressure-distribution have been obtained. Results have been illustrated graphically and compared with some earlier analyses.

Интегральный метод для расчета ламинарного течения во входном участке трубы с использованием импульса и энергии для жидкостей, которые характерисуются степеными законами, был изучен с помощью применения расширения по Волькову метода Кармана-Польгаусена. Для развития пограничного слоя и для распределения давления во входном участке трубы даны аналитические выражения. Результаты, которые сравниваются с ранными результатами, приведены в виде графиков.

1. Introduction

Flow in the inlet region of a tube drawing fluid from a large upstream reservoir is of considerable importance in many branches of engineering. Since a large variety of materials exhibit non-Newtonian behaviour, various attempts have been made to study the flow behaviour of such fluids in the inlet of a circular tube.

To study the developing laminar flow of power-law fluids in a circular tube BOGUE [1] and KAPUR and GUPTA [2] applied integral forms of equations of continuity and momentum to a model consisting of an inviscid core and a developing boundary layer, whose velocity profile is expressible in form of a polynomial.

GUPTA [3] improved the solution of [1, 2] by applying macroscopic mechanical energy balance to all the fluid in the inlet region instead of applying only to the inviscid core as has been done in [1, 2]. In [3] the analysis includes the loss of energy due to viscous dissipation in the boundary layer, which becomes increasingly significant, because as the fluid moves downstream the boundary layer — the region in which viscous effects dominate — grows thicker. The effectiveness of integral methods [1, 2, 3] is due to the fact that the values of the integrals occuring in momentum and energy integral equations do not deviate much if the integrands depart a little from their real values provided the boundary conditions are observed. The main source of error appears to be the wall shear stress term appearing in the momentum integral equation. When this term is replaced with the help of an assumed boundary layer distribution, the error introduced may be appreciable.

VOLKOV [4] has used a refined version of KARMAN-POHLHAUSEN method for investigating the boundary layer on a flat plate. The refinement consists in expressing the wall shear stress term in form of integrals involving unknown velocity distribution. He has observed that the boundary layer thickness and the plate shear stress are closer to the real values than those obtained from KARMAN-POHLHAUSEN method, and more significantly the refined values have been found to be less sensitive to the choice of the velocity distribution. Thus to obtain a better flow picture of power-law fluids in the inlet region of a circular tube in the following section we incorporate VOLKOV's suggestion in its momentum-energy-integral solution [3].

2. Momentum and Energy Integral Equations

The boundary layer equations governing steady constant property axisymmetric flow of power-law fluids which are characterised by the rheological relation

$$t_{ij} = k \left| \left(\sum_{l=1}^{3} \sum_{m=1}^{3} e_{lm} e_{ml} \right)^{1/2} \right|^{(n-1)} e_{ij} \tag{1}$$

between the components of deviatoric stress tensor t_{ij} and those of rate of strain tensor e_{ij}; k and n being the consistency and the flow behaviour index of the fluid; in the inlet region of a circular tube of radius R are

$$\frac{\partial u}{\partial x} + \frac{1}{r} \frac{\partial}{\partial r}(rv) = 0 \,, \tag{2}$$

$$u \frac{\partial u}{\partial x} + v \frac{\partial u}{\partial r} = -\frac{1}{\varrho} \frac{\mathrm{d}p}{\mathrm{d}x} + \frac{\nu}{r} \frac{\partial}{\partial r} \left(r \left| \frac{\partial u}{\partial r} \right|^{n-1} \frac{\partial u}{\partial r} \right), \tag{3}$$

with

$$\nu = \frac{k 2^{\frac{n-1}{2}}}{\varrho} = \frac{K}{\varrho} \text{(say)} \,. \tag{4}$$

Cylindrical polar coordinates x, r are used; where the x axis coincides with the axis of the pipe and r is the radial distance from the x axis and the origin is taken at the entrance section. The fluid is assumed to enter the tube with uniform velocity V and the other symbols have their usual significance.

The boundary conditions are

$$x = 0\,,\quad u = V \quad \text{for all } r\,,\quad y = R\,,\quad u = 0 \quad \text{for all } x\,,\quad y = 0\,,\quad u_r = 0 \quad \text{for all } x\,. \tag{5}$$

Equation of continuity when integrated from the tube wall $r = R$ to some r over the flow cross section gives the radial velocity at some r as

$$v|_r = \frac{1}{r}\int\limits_r^R u_x r\,\mathrm{d}r\,. \tag{6}$$

Elimination of v between (6) and the momentum equation (3) and subsequent integration over the flow cross section from arbitrary r to R gives

$$\int\limits_R^r 2uu_x r\,\mathrm{d}r - u\int\limits_R^r u_x r\,\mathrm{d}r + \frac{1}{\varrho}\frac{\mathrm{d}p}{\mathrm{d}x}\frac{1}{2}(r^2 - R^2) = \nu\big((r\,|u_r|^{n-1}\,u_r)\,|_r - R(|u_r|^{n-1}\,u_r)|_R\big)\,. \tag{7}$$

When $r = 0$ this reduces to the momentum integral equation

$$\int\limits_0^R 2\,uu_x r\,\mathrm{d}r + \frac{1}{2}\,\frac{1}{\varrho}\,\frac{\mathrm{d}p}{\mathrm{d}x}R^2 = \nu R(|u_r|^{n-1}\,u_r)|_R\,. \tag{8}$$

Again integrating (7) over the flow cross section one arrives at

$$\nu R(|u_r|^{n-1}\,u_r)\,|_R\,\frac{1}{2}\,R^2 = \int\limits_R^0\int\limits_R^r 2uu_x r\,\mathrm{d}r\,r\,\mathrm{d}r - \int\limits_R^0 u\int\limits_R^r u_x r\,\mathrm{d}r\,r\,\mathrm{d}r + \frac{R^4}{8\varrho}\frac{\mathrm{d}p}{\mathrm{d}x} - \nu\int\limits_R^0 |u_r|^{n-1}\,u_r r^2\,\mathrm{d}r\,. \tag{9}$$

Combining (8) and (9) we obtain modified momentum integral equation in the form

$$\frac{R^4}{8\varrho}\frac{\mathrm{d}p}{\mathrm{d}x} = R^2\int\limits_R^0 uu_x r\,\mathrm{d}r + \int\limits_R^0\int\limits_R^r 2uu_x r\,\mathrm{d}r\,r\,\mathrm{d}r - \int\limits_R^0 u\int\limits_R^r u_x r\,\mathrm{d}r\,r\,\mathrm{d}r + \nu\int\limits_R^0 |u_r|^{n-1}\,(-u_r)\,r^2\,\mathrm{d}r\,. \tag{10}$$

The equation describing rate of change of kinetic energy per unit mass can be obtained by multiplying each term of (3) by $2u$. This equation when integrated over the flow produces the energy integral equation as

$$-\frac{VR^2}{\varrho}\frac{\mathrm{d}p}{\mathrm{d}x} = \int\limits_0^R 3u^2u_x r\,\mathrm{d}r + 2\nu\int\limits_0^R |u_r|^{n-1}\,u_r^2 r\,\mathrm{d}r\,. \tag{11}$$

The integrated version of principle of conservation of mass may be stated as

$$Q = 2\int\limits_0^R ur\,\mathrm{d}r\,. \tag{12}$$

3. Boundary Layer Development and Pressure Drop in the Inlet Region

The axial velocity distribution is assumed to be of the form

$$\begin{aligned} u &= U \qquad (0 \leqq r \leqq R - \delta) \quad \text{for all} \quad x > 0\,,\\ u &= U\left(1 - \left(1 - \frac{R - r}{\delta}\right)^{(n+1)/n}\right) \qquad (R - \delta \leqq r \leqq R) \quad \text{for all} \quad x > 0\,. \end{aligned} \tag{13}$$

Substitution of (13) in (12) yields

$$\frac{U}{V} = \frac{1}{1 - \dfrac{2n\delta^*}{2n+1} + \dfrac{2n^2}{(2n+1)(3n+1)}\delta^{*2}} = \frac{1}{1 - A_1\delta^* + A_2\delta^{*2}}\,(\text{say})\,, \tag{14}$$

δ^* $(= \delta/R)$ denotes the dimensionless boundary layer thickness.

Substitution of (13) in (10) and (11) and subsequent elimination of $\mathrm{d}p/\mathrm{d}x$ between them results in

$$x^* = \frac{x/R}{N_{Re}} = \frac{1}{64}\left(\frac{3n+1}{n+1}\right)^n \int\limits_0^{\delta^*} \frac{\delta^{*n}(1 - A_1\delta^* + A_2\delta^{*2})^{n-3}\,J(\delta^*)\,\mathrm{d}\delta^*}{\dfrac{n+1}{4n+2}\left(1 - \dfrac{n\delta^*}{3n+1}\right) - (1 - A_1\delta^* + A_2\delta^{*2})\,\delta^*\left(1 - \dfrac{2}{3}\delta^* + \dfrac{1}{6}\delta^{*2}\right)}\,, \tag{15}$$

with

$$\begin{aligned} J(\delta^*) = {}& (1 - A_1\delta^* + A_2\delta^{*2})^2\,(4(-B_1 + 2B_2\delta^*) - 16G(\delta^*)) +\\ &+ (1 - A_1\delta^* + A_2\delta^{*2})\,(16(-A_1 + 2A_2\delta^*)\,F(\delta^*) - 8(-A_1 + 2A_2\delta^*)\,(1 - B_1\delta^* + B_2\delta^{*2}) +\\ &+ (C_1 - 2C_2\delta^*)) + 3(-A_1 + 2A_2\delta^*)\,(1 - C_1\delta^* + C_2\delta^{*2})\,, \end{aligned} \tag{16}$$

where

$$F(\delta^*) = \frac{1}{8} - \frac{n(5n+4)\,\delta^*}{2(2n+1)(3n+2)} + \frac{n^2(39n^2+53n+17)\,\delta^{*2}}{2\,(2n+1)^2(3n+1)(3n+2)} - \frac{n^3(228n^3+407n^2+229n+41)\,\delta^{*3}}{(2n+1)^2(3n+1)(4n+1)(3n+2)(5n+2)} + \frac{n^4(1500n^4+3187n^3+2404n^2+770n+89)\,\delta^{*4}}{2(2n+1)^2(3n+2)^2(4n+1)(5n+1)(3n+2)(5n+2)},$$

$$G(\delta^*) = -\frac{n(n+1)}{2(2n+1)(3n+2)} + \frac{n^2(15n^2+23n+8)\,\delta^*}{2(2n+1)^2(3n+1)(3n+2)} - \frac{n^3(792n^3+1539n^2+918n+171)\,\delta^{*2}}{6(2n+1)^2(3n+1)(4n+1)(3n+2)(5n+2)} + \frac{n^4(2400n^4+5464n^3+4330n^2+1436n+170)\,\delta^{*3}}{4(2n+1)^2(3n+1)^2(4n+1)(5n+1)(3n+2)(5n+2)},$$

$$B_1 = \frac{2n(4n+3)}{(3n+2)(2n+1)}, \quad B_2 = \frac{n^2(9n+7)}{(2n+1)(3n+2)(3n+1)}, \quad C_1 = \frac{2n(18n^2+28n+11)}{(2n+1)(3n+2)(4n+3)},$$

$$C_2 = \frac{n^2(216n^3+477n^2+350n+85)}{(2n+1)(3n+2)(4n+3)(3n+1)(5n+3)},$$

and N_{Re} $(= 8\varrho R^n V^{2-n} n^n / K(3n+1)^n)$ is the abbreviation for the Reynolds number.

This relation determines the boundary layer thickness at a given station. Boundary layer development has been displayed in Figs. 1 (a) and 1 (b) for pseudoplastic ($n = 0.75$) and dilatant ($n = 1.25$) fluids. Inlet region pressure distribution can be obtained from either (10) or (11). If p_0 is the pressure at the entry and p is the pressure at a distance x from the entrance plane then

$$p^* = \frac{p_0 - p}{\frac{1}{2}\varrho V^2} = \frac{1 - C_1\delta^* + C_2\delta^{*2}}{(1 - A_1\delta^* + A_2\delta^{*2})^3} - 1 + \int\limits_0^{\delta^*} \frac{32\left(\frac{n+1}{3n+1}\right)^n \left(\frac{n+1}{2n+1}\right)\left(1 - \frac{n\delta^*}{3n+1}\right)\frac{dx^*}{d\delta^*}\,d\delta^*}{(1 - A_1\delta^* + A_2\delta^{*2})^{n+1}\,\delta^{*n}}. \quad (17)$$

The slope of the pressure drop-distance curve is continuously changing in the inlet region. When the flow becomes fully established the slope of $p^* - 32x^*$ curve is unity. And that just before the flow becomes fully developed can be found from (15) and (17). It is given by

$$\tan\theta = \frac{((n+1)/(3n+1))^n}{(1 - A_1 + A_2)^{n+1}}\left[\frac{n+1}{3n+1} + \frac{(2C_2 - C_1)(1 - A_1 + A_2) + (1 - C_1 + C_2)(3A_1 - 6A_2)}{64J(1)} \times \right.$$
$$\left. \times \left(\frac{n+1}{3n+1} - (1 - A_1 + A_2)\right)\right] = 1 \text{ for all } n\,. \quad (18)$$

The axial velocity distribution in the boundary layer is determined by combining (13) and (14) and is given by

$$u^* = \frac{u}{V} = \frac{1}{1 - A_1\delta^* + A_2\delta^{*2}}\left(1 - \left(1 - \frac{R-r}{\delta}\right)^{(n+1)/n}\right). \quad (19)$$

This distribution at various radial distances is shown in Fig. 3.

4. Results

Fargie and Martin [5] have described a method of predicting the laminar flow development in a circular pipe, which consists in employing both the integral and differential forms of the momentum equation. This analysis differs from that of Campbell and Slattery [6] in the determination of the pressure gradient; in [6] dp/dx has been determined by an application of mechanical energy balance to all the fluid in the inlet region, whereas in [5] the normalised reciprocal pressure gradient has been assumed to be a quadratic polynomial of nondimensional inviscid core thickness.

Mashelkar [7] has extended the procedure of [5] to obtain the hydrodynamic entrance region flow of pseudoplastic fluids in a circular pipe. Continuation of Mashelkar's work for dilatant fluids has been presented in the Appendix at the end of this paper. Discontinuous curves (— — — —) in the figures represent these results, while those with dots (—.—.—.—) depict the recalculated results of [3] for the assumed axial velocity distribution [13].

Boundary layer thickness as a function of location is shown in Fig. 1. For pseudoplastic ($n = 0.75$) fluids, analysis of Mashelkar [7] predicts the most rapid rate of growth of boundary layer thickness. As compared to the predictions based on [3], the present theory predicts that the growth of boundary layer thickness is less rapid when $\delta^* \leq 0.7$ and more rapid when $\delta^* \geq 0.8$. The transition occurs when δ^* lies between 0.7 and 0.8. This in turn implies that the length of the inlet region as predicted by [7], present analysis and those based on [3] are in increasing order of magnitude.

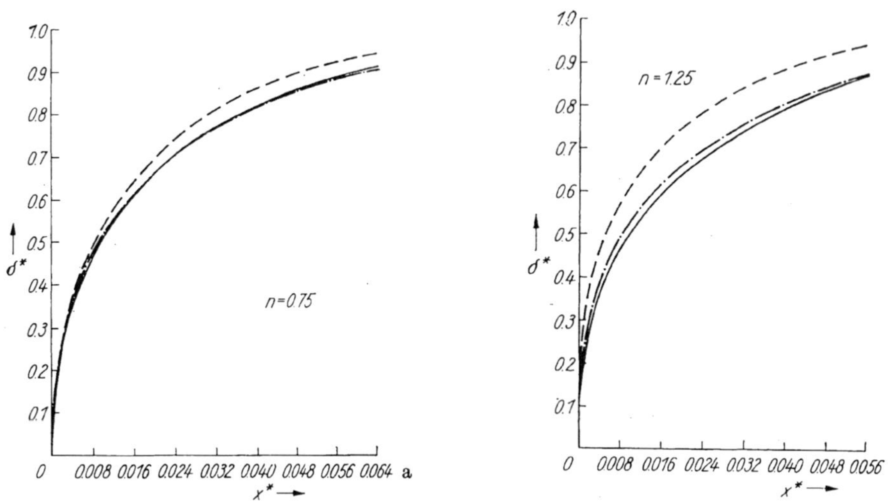

Fig. 1. Boundary layer development in the inlet region. (a) for pseudoplastic fluids; (b) for dilatant fluids

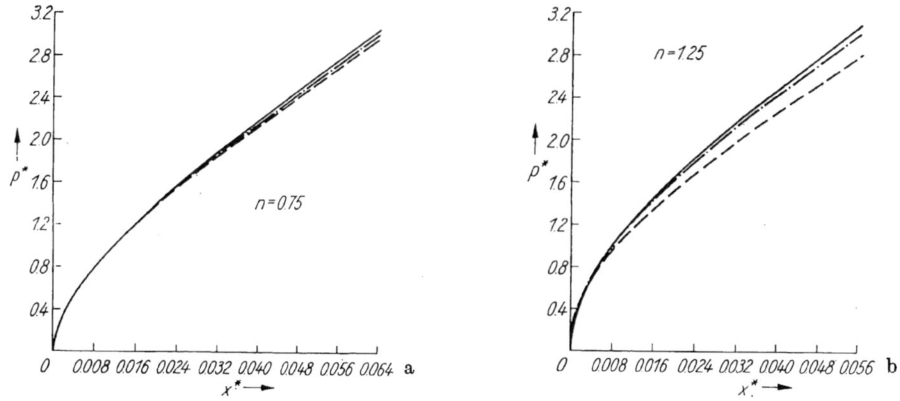

Fig. 2. Axial pressure distribution in the inlet region. (a) for pseudoplastic fluids; (b) for dilatant fluids

For dilatant fluids ($n = 1.25$) the station at which a given δ^* is attained is seen to be most inward for the present analysis and closest to the entry plane as predicted on the basis of Fargie and Martin's theory. The findings based on [3] lie between these two extremes. Dimensionless inlet region pressure distribution has been displayed in Fig. 2. Both for pseudoplastic and dilatant fluids, except possibly near the entry, at a given location the pressure drop as obtained from this work, [3] and [7] are in descending order of magnitude. In fully developed state the slope of $p^* - 32x^*$ curve is unity. In the developing region this slope continuously changes and the same at the end of the developing region is found to be unity for all values of n. The present theory thus predicts that the slope of the pressure drop-distance curve is continuous in the entire flow field. Other analyses have, on the other hand, predicted discontinuity in the slope of $p^* - 32x^*$ curve.

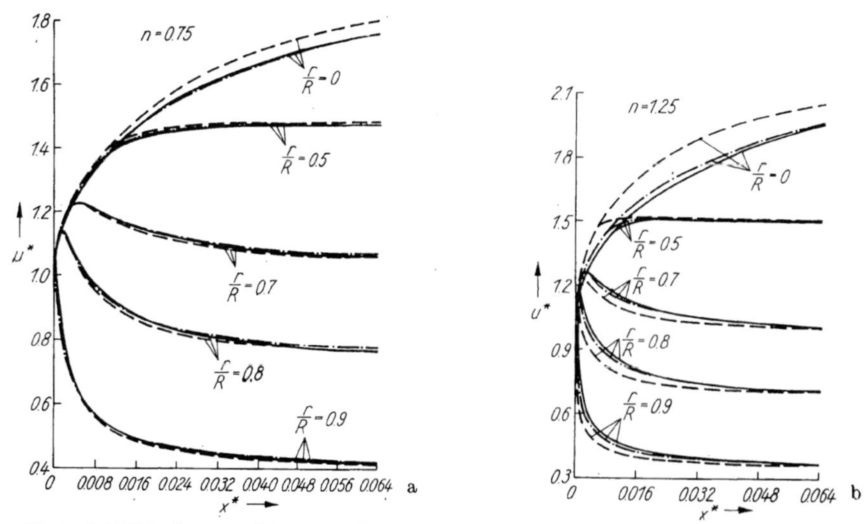

Fig. 3. Axial Velocity versus distance from the entry plane (a). for pseudoplastic fluids; (b) for dilatant fluids

Fig. 3 compares the axial velocity distribution for the three theories under consideration. This analysis reduces to that of [8] for Newtonian fluids.

5. Concluding Remarks

VOLKOV's refinement of KARMAN-POHLHAUSEN method has been incorporated in the momentum energy integral solution [3] for the laminar developing flow of power-law fluids in a circular pipe. The assumed axial velocity distribution reduces to the one in the fully established state when the boundary layer merges with the axis of the pipe.

The analysis has shown that the slope of the pressure drop-distance curve is continuous at the beginning of the fully developed state. This appears to be a significant achievement of this investigation. In view of all this it is felt that the results presented in this paper give the best over all picture of the power-law fluid flow in the inlet region of a circular pipe.

Use of computing facilities provided by the University of Dar es Salaam are gratefully acknowledged.

Appendix

Here we continue MASHELKAR's analysis [7] and present below analytical expressions for the flow of dilatant fluids in the inlet region of a circular pipe.

For dilatant fluids the normalised inverse pressure gradient when plotted as a function of nondimensional core thickness does not show any point of inflection and, therefore, pressure can be determined from equation (24) of [7].

Combination of equations (24) and (14) of [7] after simplification gives

$$x = \int_1^\alpha \frac{(2n+1)(3n+1)}{n(3n+2)} \left(\frac{2}{f_1 f_3}\right)^n \frac{H(\alpha)(1-\alpha)^{n+1}\,d\alpha}{\alpha(f_2+f_3\alpha+2\alpha^2)^{3-n}}, \tag{A.1}$$

$$p = \int_1^\alpha \frac{(2n+1)(3n+1)(\alpha-2)\,H(\alpha)\,d\alpha}{n(3n+2)\,\alpha(f_2+f_3\alpha+2\alpha^2)^3}, \tag{A.2}$$

with

$$H(\alpha) = (f_5 + f_6 2\alpha)(f_2 + f_3\alpha + 2\alpha^2) - 2(f_3 + 4\alpha)(f_4 + f_5\alpha + f_6\alpha^2)$$

and other symbols are the same as in [7].

References

1 BOGUE, D. C., Entrance effects and prediction of turbulence in non-Newtonian fluids, Ind. Engg. Chem. **51**, pp. 874–878 (1959).
2 KAPUR, J. N., and GUPTA, R. C., Power-law fluid flow in the inlet region of a circular pipe, The Maths. Seminar **3**, pp. 55–67 (1963).
3 GUPTA, R. C., Flow behaviour of power-law fluids in the entry region of a circular pipe, J. de Mécanique **8**, pp. 207–220 (1969).
4 VOLKOV, V. N., A refinement of the KARMAN-POHLHAUSEN integral method in boundary layer theory, J. Engg. Phys. **9**, pp. 371 to 374 (1965).
5 FARGIE, D., and MARTIN, B. W., Developing laminar flow in a pipe of circular cross section, Proc. Roy. Soc. **231** A, pp. 461–476 (1971).
6 CAMPBELL, W. D., and SLATTERY, J. C., Flow in the entrance of a tube, J. Basic Engg. (Trans. ASME) **85** D, pp. 41–46 (1963).
7 MASHELKAR, R. A., Hydrodynamic entrance region flow of pseudoplastic fluids, Proc. Instn. Mech. Engrs. **188**, pp. 683–689 (1975).
8 GUPTA, R. C., Laminar flow in the entrance of a tube, Appl. Sci. Res. **33**, pp. 1–10 (1977).

Eingereicht am 15. 5. 1980

Anschrift: Dr. R. C. GUPTA, University of Dar es Salaam, Department of Mathematics, P.O. Box 35062, Dar es Salaam, Tanzania

ZAMM 61, 305–314 (1981)

H. MURAKAMI / A. MAEWAL / G. A. HEGEMIER

A Mixture Theory for Heat Conduction in Laminated Composites

Die asymptotische Methode der Mehrfachmaßstäbe wird benutzt, um eine Kontinuumstheorie mit Mikrostruktur für die Wärmeleitung eines periodisch laminierten Mediums aufzustellen. Die sich hieraus ergebende Theorie hat die Form einer homogenen binären Mischungstheorie mit einem gemeinsamen Mikrotemperaturgradienten, der normal zur Grenzfläche orientiert ist. Das Modell enthält drei Erhaltungsgleichungen, die die mittlere Temperatur zweier Komponenten und den Mikrotemperaturgradienten enthalten.
Die Mischungsgleichungen werden auch auf die Wärmeleitung in einem schief laminierten Halbraum, dessen freie Oberfläche plötzlich erwärmt wird, angewendet. Der Einfluß des Laminatwinkels wird geklärt.

An asymptotic method of multiple scales is used to construct a continuum theory with microstructure for heat conduction of a periodically laminated medium. The resulting theory is in the form a homogeneous binary mixture theory with a common microtemperature gradient oriented normal to the interfaces. The model contains three conservation equations which include averaged temperatures of two constituents and microtemperature gradient. Finally, the mixture equations are applied for heat conduction in obliquely laminated half space which is suddenly heated on its free surface, and the effect of lay-up angle is clarified.

Асимптотический метод многократных масштабов используется для получения теории сплошной среды с микроструктурой для темлопроводности периодически ламинаризованной жидкости. Теория, полученная таким путем, можно записать в виде однородной бинарной теории смешивания с одним градиентом микротемпературы, направленным перпендикулярно к поверхности среды. Модель состоит из трех уравнений сохранения относительно средней температуры двух компонент и градиента микротемпературы. Уравнения смешивания применяются для теплопроводности в косеугольно ламинизованном полупространстве, свободная поверхность которого мгновенно нагревается. Выясняется эффект угла ламинаризации.

1. Introduction

For unsteady state heat conduction in periodically laminated media it has been found that a model which utilizes an effective thermal conductivity or an effective thermal diffusivity constitutes a steady state approximation, i.e., zero frequency limit approximation either by an asymptotic model construction [1] or by solving exact characteristic equations for time-harmonic heat flow in the direction parallel to the layering [2]. It has also been revealed that the effective thermal conductivity theory may not be accurate at high frequencies due to dispersion phenomena in laminated composites, which are induced by the influence of microstructure of the composite.

In order to incorporate the dispersive phenomena the differential-difference method has been applied to construct a higher order model for heat conduction in laminated composites [1].

For heat conduction in the direction parallel to the layering NAYFEH [3], [4] has applied the mixture theory which has been developed to model wave propagation in laminated and fiber reinforced composites [5]. However, the application of the mixture theory was only possible to the wave-guide type problems.

In this paper a binary mixture theory which is free from such a wave-guide limitation is developed for heat conduction in laminated composites with periodic microstructure. The construction of the theory is based upon the method of multi-variable asymptotic expansions developed to model mechanical behavior of laminated composites [6], [7]. This method is suggested by the fact that, along a direction normal to the laminae, there are two length scales over which significant variations in temperature and heat flux profiles occur, and these scales differ by at least an order of magnitude in most problems of practical interest. The resulting equations consist of three equations for averaged temperature $T^{(1a)}$, $T^{(2a)}$ and microtemperature gradient S as shown in equations (50–52).

In order to infer the accuracy of the theory to model unsteady state heat conduction, time decay factors of spacially harmonic temperature fields are calculated for the first mode using the mixture equations and results are compared with the exact ones. Good agreement between two results is observed even down to wavelengths equal to twice the typical composite microdimension. This implies that the theory constitutes an attractive alternative to direct numerical solutions, for problems where the energy is partitioned primarily in the first mode, and where dominant signal wavelength are not less than the typical material microdimension.

As an illustrative example, the mixture equations are applied for heat conduction in obliquely laminated half space which is suddenly heated on its free surface.

2. Formulation

Consider a domain $\bar{D}$ which consists of a large number of two alternating homogeneous laminae which are identified as "material 1" and "material 2" as shown in Fig. 1. It is assumed that there is no thermal resistance across the interfaces of the two constituents. A typical cell which represents the geometrical microstructure of the composite is shown in Fig. 2. Let a rectangular Cartesian coordinate system $\bar{x}_1$, $\bar{x}_2$, $\bar{x}_3$ be selected with $\bar{x}_3$ normal to the middle plane of the central cell lamina and $\bar{x}_2$, $\bar{x}_3$ in this plane.

For notational convenience forms $(\ )^{(\alpha)}$, $\alpha = 1, 2$ will denote quantities associated with material α. The usual Cartesian indicial notation is employed where Latin indices range from 1 to 3 and repeated indices imply summation convention, unless otherwise noted. In addition, the notations $(\ )_{,i} = \partial(\ )/\partial\bar{x}_i$ and $(\dot{\ }) = \partial(\ )\ \partial/\bar{t}$ shall be employed where $\bar{t}$ is time. Quantities of the form $(\bar{\ })$ and $(\ )$ denote dimensional and nondimensional variables, respectively.

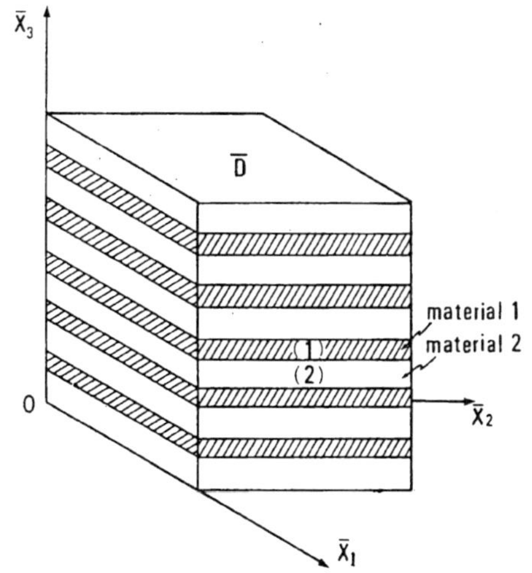

Fig. 1. Geometry and Coordinate System

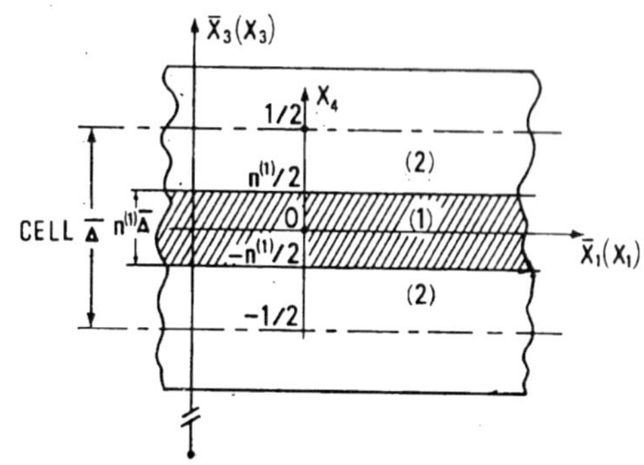

Fig. 2. Unit Cell

With the aid of the forgoing notation, the basic relations for the temperature $\bar{T}^{(\alpha)}$ and the heat flux vector $\bar{Q}_i^{(\alpha)}$ in the two constituents are:

(a) Conservation of Energy:

$$\bar{Q}_{i,i}^{(\alpha)} = -\bar{\mu}^{(\alpha)}\dot{\bar{T}}^{(\alpha)} , \tag{1}$$

where $\bar{\mu}^{(\alpha)}$ is heat capacity;

(b) Fourier's Law of Heat Conduction:

$$\bar{Q}_i^{(\alpha)} = -\bar{k}^{(\alpha)}\bar{T}_{,i}^{(\alpha)} \tag{2}$$

where $\bar{k}^{(\alpha)}$ is thermal conductivity;

(c) Interface Conditions:

$$\bar{T}^{(1)} = \bar{T}^{(2)} , \qquad \bar{Q}_3^{(1)} = \bar{Q}_3^{(2)} \quad \text{on interfaces;} \tag{3}$$

(d) Initial Condition at $\bar{t} = 0$ and Appropriate Boundary Data on the Boundary of $\bar{D}$.

Conditions (a—d) pose an initial boundary value problem in which the continuity conditions are to be satisfied at a multitude of interfaces. As a result of the discrete character of the composite it is very difficult to obtain a direct solution of this problem. The objective of the subsequent analysis is to alleviate such difficulties by deriving a set of partial differential equations with constant coefficients whose solution can be used to approximate the solution of the problem posed by conditions (a—d). This effort is facilitated by a judicious scaling of both dependent and independent variables. For this purpose the following quantities are introduced:

$\bar{\Lambda}$: typical macrodimension;
$\bar{\Delta}$: composite cell length;
$\bar{t}_{(m)} = \bar{\mu}_{(m)}\bar{\Lambda}^2/\bar{k}_{(m)}$: characteristic thermal diffusion time;
$\bar{k}_{(m)}, \bar{\mu}_{(m)}$: reference thermal conductivity and heat capacity;
$\bar{T}_0$: reference temperature;
$\bar{Q}_{(m)} = \bar{k}_{(m)}\bar{T}_0/\bar{\Lambda}$: reference heat flux;
$\varepsilon = \bar{\Delta}/\bar{\Lambda}$: ratio of micro-to-macrodimensions.

With the aid of forgoing notation, the following nondimensional quantities are defined:

$$\begin{aligned} &(x_1, x_2, x_3) = (\bar{x}_1, \bar{x}_2, \bar{x}_3)/\bar{\Lambda} , \qquad t = \bar{t}/\bar{t}_{(m)} , \\ &T^{(\alpha)} = (\bar{T}^{(\alpha)} - \bar{T}_0)/\bar{T}_0 , \qquad Q_i^{(\alpha)} = \bar{Q}_i^{(\alpha)}/\bar{Q}_{(m)} , \\ &k^{(\alpha)} = \bar{k}^{(\alpha)}/\bar{k}_{(m)} , \qquad \mu^{(\alpha)} = \bar{\mu}^{(\alpha)}\bar{\mu}/_{(m)} . \end{aligned} \tag{4}$$

Observing that there are two length scales in x_3-direction, one is of the order of the macrodimension which is unity in the scaled system and the other is the order of the cell dimension ε, a multi-variable expansion procedure is adopted for smoothing of the discrete media where $x_4 = x_3/\varepsilon$ is introduced as a new independent variable [6], [7]. All field variables are now functions of x_4 as well as the macrovariables x_i $(i = 1, 3)$, and time t, i.e.,

$$f(x_1, x_2, x_3, t; \varepsilon) = f^*(x_1, x_2, x_3, x_4, t; \varepsilon) . \tag{5}$$

For any field variable f, one obtains

$$\frac{\partial}{\partial x_3} f(x_1, x_2, x_3, t; \varepsilon) = \frac{\partial}{\partial x_3} f^*(x_1, x_2, x_3, x_4, t; \varepsilon) + \frac{1}{\varepsilon}\frac{\partial}{\partial x_4} f^*(x_1, x_2, x_3, x_4, t; \varepsilon) . \tag{6}$$

In the sequel * will be dropped for simplicity.

On introduction of x_i, x_4, and the nondimensional variables defined in (4), equations (1—3) take the form

$$Q^{(\alpha)}_{i,i} + \frac{1}{\varepsilon} Q^{(\alpha)}_{3,4} = -\mu^{(\alpha)} \dot{T}^{(\alpha)} , \qquad Q^{(\alpha)}_i = -k^{(\alpha)}\left(T^{(\alpha)}_{,i} + \delta_{i3} \frac{1}{\varepsilon} T^{(\alpha)}_{,4}\right), \tag{7), (8}$$

$$T^{(1)} = T^{(2)} , \qquad Q^{(1)}_3 = Q^{(2)}_3 \quad \text{on interfaces} , \tag{9}$$

where the notations $(\ )_{,i} = \partial(\ )/\partial x_i$, $(\dot{\ }) = \partial(\ )/\partial t$ have been utilized and δ_{ij} is the KRONECKER delta which is equal to 1 when $i = j$ and zero otherwise.

Equations (7—9) can be viewed as a set of partial differential equations in x_i, x_4, and t, in which coefficients are independent of x_i, t, but are periodic functions of x_4 with a unit period. If the domain of definition of (7—9) was such that $-\infty < x_4 < \infty$, it would follow from FLOQUET Theory [8] that there exists a set of solutions of (7—9) which are periodic in x_4 with unit period length. As in [6], [7] this periodicity condition is adopted in the following analysis limiting the applicability of the resulting theory to the domain where the length of it in the x_3-direction is sufficiently large.

As a result of this periodicity condition, it is necessary to treat only a typical unit cell associated with the x_4-direction, as shown in Fig. 2, where the domain $A^{(\alpha)}$ occupied by material α is defined as

$$A^{(1)} = \left[x_4 \colon |x_4| < \frac{n^{(1)}}{2}\right], \quad A^{(2)} = \left[x_4 : \frac{n^{(1)}}{2} < |x_4| < \frac{1}{2}\right]. \tag{10}$$

In definitions (10), $n^{(\alpha)}$ is the volume fraction of material α and satisfies

$$n^{(1)} + n^{(2)} = 1 . \tag{11}$$

With the above choice of the unit cell, the interface conditions (9) must be satisfied at $x_4 = \pm n^{(1)}/2$ and the periodicity conditions are given by

$$f\left(x_i, -\frac{1}{2}, t; \varepsilon\right) = f\left(x_i, \frac{1}{2}, t; \varepsilon\right), \tag{12}$$

for f represents any field variable.

The formulation of the problem is now complete. The key elements introduced so far are 1) the microcoordinate x_4 and 2) the periodicity of the field variables in this microcoordinate.

3. Mixture Equations

Binary mixture equations for the composite can be obtained by performing an averaging operation according to

$$f^{(\alpha a)} = \frac{1}{n^{(\alpha)}} \int\limits_{A(\alpha)} f^{(\alpha)}(x_i, x_4, t; \varepsilon)\, dx_4 . \tag{13}$$

Upon averaging (7) with use of the periodicity condition for the heat flux Q_3, one obtains the mixture law of conservation of energy:

$$Q^{(\alpha p)}_{i,i} - (-1)^{\alpha} P = -\mu^{(\alpha p)} \dot{T}^{(\alpha a)} \tag{14}$$

where partial heat flux and heat capacity have been introduced according to

$$Q^{(\alpha p)}_i = n^{(\alpha)} Q^{(\alpha a)}_i , \qquad \mu^{(\alpha p)} = n^{(\alpha)} \mu^{(\alpha)} . \tag{15}$$

The variable P is an interaction term reflecting heat transfer between the two constituents; this is defined in terms of the interface heat flux given by

$$P \equiv \frac{1}{\varepsilon} \left\{ Q^{(\alpha)}_3 \bigg|_{x_4 = \frac{n^{(1)}}{2}} - Q^{(\alpha)}_3 \bigg|_{x_4 = -\frac{n^{(1)}}{2}} \right\} . \tag{16}$$

Upon averaging (8) and applying periodicity condition one obtains mixture conduction law:

$$Q^{(\alpha p)}_i = -k^{(\alpha p)} T^{(\alpha a)}_{,i} + (-1)^{\alpha}\, \delta_{i3} k^{(\alpha)} \mathrm{m}\, S \tag{17a}$$

where

$$k^{(\alpha p)} = n^{(\alpha)} k^{(\alpha)} \tag{17b}$$

and

$$S \equiv \frac{1}{\varepsilon} \left\{ T^{(\alpha)} \bigg|_{x_4 = \frac{n^{(1)}}{2}} - T^{(\alpha)} \bigg|_{x_4 = -\frac{n^{(1)}}{2}} \right\} . \tag{18}$$

The quantity S is an interaction term associated with the conduction law and shall be termed "microtemperature gradient". In the subsequent analysis the microtemperature gradient will be treated as dependent variable in addition to the average temperature $T^{(\alpha a)}$.

In order to "close" the mixture relations, it is necessary to construct (a) appropriate conservation equation associated with the microtemperature gradient and (b) constitutive equations for P and for the heat flux-type quantities associated with the microtemperature gradient. For this purpose an asymptotic procedure will be utilized which is an extension of the regular asymptotic method [9].

4. Asymptotic Analysis

The premise that the ratio of micro- to macrodimensions of the composite is small compared to unity, i.e., $\varepsilon \ll 1$ and the form of the scaled equations (7—9) suggests the following asymptotic expansions for all dependent variables:

$$\{T^{(\alpha)}, Q_i^{(\alpha)}\}\,(x_k, x_4, t; \varepsilon) = \sum_{n=0}^{\infty} \varepsilon^n \{T_{(n)}^{(\alpha)}, Q_{i(n)}^{(\alpha)}\}\,(x_k, x_4, t)\,. \tag{19}$$

If equations (19) are substituted into (7—9) and terms of similar ε-order are equated in the usual manner, a sequence of problems is obtained. The first of the equation in this sequence furnishes

$$Q_{3(0),4}^{(\alpha)} = T_{(0),4}^{(\alpha)} = 0\,. \tag{20}$$

Thus the quantities $Q_{3(0)}^{(\alpha)}$, $T_{(0)}^{(\alpha)}$ are independent of the microcoordinate x_4, i.e.,

$$\{T^{(\alpha)}, Q_3^{(\alpha)}\}_{(0)} = \{T, Q_3\}_{(0)}\,(x_1, x_2, x_3, t) \tag{21}$$

where zeroth order expansions of the interface conditions (9) have also been utilized. The remaining system of equations are

$$Q_{i(n),i}^{(\alpha)} + Q_{3(n+1),4}^{(\alpha)} = -\mu^{(\alpha)}\dot{T}_{(n)}^{(\alpha)}\,, \qquad n \geqq 0\,, \tag{22}$$

$$Q_{i(n)}^{(\alpha)} = -k^{(\alpha)}\left(T_{(n),i}^{(\alpha)} + \delta_{i3} T_{(n+1),4}^{(\alpha)}\right), \qquad n \geqq 0\,. \tag{23}$$

Because of (21) the higher order terms in (19) must satisfy normalization condition; a convenient choice is

$$T_{(n)}^{(1)} = Q_{i(n)}^{(1)} = 0 \quad \text{on} \quad x_4 = 0\,, \qquad n \geqq 1\,. \tag{24}$$

Equations (21), (22) and (23) imply that for $n \geqq 1$

$$T_{(2n)}^{(\alpha)} \text{ and } Q_{3(2n)}^{(\alpha)} \text{ are even functions of } x_4\,, \tag{25a}$$

$$T_{(2n-1)}^{(\alpha)} \text{ and } Q_{3(2n-1)}^{(\alpha)} \text{ are odd functions of } x_4\,. \tag{25b}$$

These properties follow from the choice of the unit cell, i.e., from the symmetrical distribution of material properties within it. In view of relations (25) and the periodicity condition, one also obtains

$$T_{(2n-1)}^{(2)} = Q_{3(2n-1)}^{(2)} = 0 \quad \text{on} \quad x_4 = \pm\frac{1}{2}\,, \qquad n \geqq 1\,. \tag{26}$$

To be added to the above are the interface conditions obtained from the expansion of (9):

$$T_{(n)}^{(1)} = T_{(n)}^{(2)}\,, \qquad Q_{3(n)}^{(1)} = Q_{3(n)}^{(2)} \quad \text{on} \quad x_4 = \pm\frac{n^{(1)}}{2}\,. \tag{27}$$

4. First Order Temperature Field and Conservation Equation forMicrotemperature Gradient

Equations (21) and (23) with $n = 0$ furnish

$$T_{(1),4}^{(\alpha)} = B^{(\alpha)}(x_1, x_2, x_3, t)\,. \tag{28}$$

The function $B^{(\alpha)}$ is related to the microtemperature gradient by (18) if $T^{(\alpha)}$ is represented by terms up to and including $O(\varepsilon^2)$ in (19).

Employing (24) with $n = 1$, (25b), (26) with $n = 1$, equation (28) can be integrated to give

$$T_{(1)}^{(\alpha)} = S(x_i, t)\, g^{(\alpha)}(x_4) \tag{29}$$

where

$$g^{(1)}(x_4) = x_4/n^{(1)}\,, \qquad g^{(2)}(x_4) = \{|x_4|/(2x_4) - x_4\}/n^{(2)}\,. \tag{30}$$

The introduction of the dependent variable S requires an additional conservation relation and this is obtained by taking g-moment of (22) with $n = 1$ and

$$g(x_4) = \begin{cases} g^{(1)}(x_4) & \text{on } A^{(1)}\,, \\ g^{(2)}(x_4) & \text{on } A^{(2)}\,. \end{cases} \tag{31}$$

Thus from (30), (31) and (22) one obtains

$$M_{i,i} + R = -\mu_{(m)}\dot{S}/12 \tag{32}$$

where

$$M_i = \sum_{\alpha=1}^{2} M_i^{(\alpha)} = \sum_{\alpha=1}^{2} \int_{A^{(\alpha)}} g^{(\alpha)}(x_4)\, Q_{i(1)}^{(\alpha)}\, dx_4 \,, \tag{33a}$$

$$R = \sum_{\alpha=1}^{2} R^{(\alpha)} = \sum_{\alpha=1}^{2} \int_{A^{(\alpha)}} g^{(\alpha)}(x_4)\, Q_{3(2),4}^{(\alpha)}\, dx_4 \,, \qquad \mu_{(m)} = \mu^{(1p)} + \mu^{(2p)} \,. \tag{33b), (33c}$$

Resulting mixture equations (14), (17), and (32) have to be accompanied by appropriate mixture constitutive relations for P, M_i and R. These are obtained systematically in the sequel by retaining terms up to and including $O(\varepsilon^2)$ in the temperature and heat flux profiles.

6. Mixture Theory Based on two Lowest Order Systems

6.1 Constitutive Relation for Interaction Term P

Equation (22) with $n = 0$, together with (21) becomes

$$Q_{3(1),4}^{(\alpha)} = C^{(\alpha)}(x_k, t) \,. \tag{34}$$

Integration of (34) over $A^{(\alpha)}$ with the use of (16), (25), and (26) yields

$$P = n^{(1)}C^{(1)} = -n^{(2)}C^{(2)} \,. \tag{35}$$

Substituting (35) into (34), one obtains

$$Q_{3(1),4}^{(\alpha)} = (-1)^{\alpha+1}\, P/n^{(\alpha)} \,. \tag{36}$$

Moreover, the conduction law (23) with $n = 1$ and equation (29) furnish

$$Q_{3(1)}^{(\alpha)} = -k^{(\alpha)}(S_{,3}\, g^{(\alpha)}(x_4) + T_{(2),4}^{(\alpha)}) \,. \tag{37}$$

Equations (36) and (37) define a boundary value problem for $T_{(2)}^{(\alpha)}$ on the cell with the appropriate boundary conditions; these are equations (24b) and (26b) with $n = 1$ and the interface continuity conditions (27).

Due to the form of forcing terms P, $S_{,3}$ the solution may be obtained in the form

$$T_{(2)}^{(\alpha)} = P v^{(\alpha)}(x_4) + S_{,3}\, w^{(\alpha)}(x_4) \,. \tag{38}$$

The differential equations, boundary, and interface conditions for $v^{(\alpha)}$, $w^{(\alpha)}$ define so-called "microstructure boundary value problems" (MBVPs). These are obtained by substituting (38) into (36) and (37), and the appropriate interface and boundary conditions (24), (26), and (27).

Since the MBVPs are elementary, only the solutions are presented below:

$$v^{(\alpha)} = -q^{(\alpha)}(x_4)/(n^{(\alpha)}k^{(\alpha)}) - \delta_{\alpha 2}\, n^{(1)}/(8k^{(1)}) \,,$$

$$w^{(\alpha)} = -\, q^{(\alpha)}(x_4)/n^{(\alpha)} - \delta_{\alpha 2}\, n^{(1)}/8 \tag{39b}$$

where

$$q^{(1)}(x_4) = x_4^2/2 \,, \qquad q^{(2)}(x_4) = (|x_4| - x_4^2 - n^{(1)}/2 + n^{(1)^2}/4)/2 \,. \tag{39c}$$

As a result, the temperature profile can be written as

$$T^{(\alpha)} = T_{(0)}(x_i, t) + \varepsilon S(x_i, t) \cdot g^{(\alpha)}(x_4) + \varepsilon^2 (P v^{(\alpha)}(x_4) + S_{,3}\, w^{(\alpha)}(x_4)) + o(\varepsilon^2) \,. \tag{40}$$

Taking the average of (40), and eliminating $T_{(0)}$, the constitutive relation for P is obtained:

$$P = \zeta (T^{(1a)} - T^{(2a)})/\varepsilon^2 - \beta S_{,3} \tag{41}$$

where

$$\zeta = 1/(v^{(1a)} - v^{(2a)}) = 12 k_{(n)} \,, \qquad \beta = (w^{(1a)} - w^{(2a)}) \cdot \zeta = k_{(n)} \,, \tag{42a), (42b}$$

and

$$k_{(n)} = 1/(n^{(1)}/k^{(1)} + n^{(2)}/k^{(2)}) \,. \tag{42c}$$

6.2 Additional Constitutive Relations

If the solutions of the MBVPs are used in (23) to determine the first order heat flux profile $Q_{i(1)}^{(\alpha)}$, then the moment defined by (33a) can be obtained from (37) and (38):

$$M_i = \sum_{\alpha=1}^{2} M_i^{(\alpha)} = -\sum_{\alpha=1}^{2} k^{(\alpha p)} S_{,i}/12 = -k_{(m)} S_{,i}/12 \,, \qquad i = 1, 2 \tag{43a}$$

$$M_3 = \sum_{\alpha=1}^{2} M_3^{(\alpha)} = \sum_{\alpha=1}^{2} n^{(\alpha)} P/12 = P/12 \tag{43b}$$

where

$$k_{(m)} = k^{(1p)} + k^{(2p)} . \tag{43c}$$

It now remains to obtain a constitutive relation for R. For this purpose equation (22) with $n = 1$ is considered; one observes that it can be written in the form

$$Q^{(\alpha)}_{3(2),4} = D^{(\alpha)}(x_\iota, t) \cdot g^{(\alpha)}(x_4) . \tag{44}$$

The function $D^{(\alpha)}$, which is independent of the microcoordinate, can be related to $R^{(\alpha)}$ by taking the $g^{(\alpha)}$-moment of (44); this procedure yields

$$D^{(\alpha)} = 12R^{(\alpha)}/n^{(\alpha)} . \tag{45}$$

Hence equation (44) reduces to

$$Q^{(\alpha)}_{3(2),4} = 12R^{(\alpha)}g^{(\alpha)}(x_4)/n^{(\alpha)} . \tag{46}$$

The boundary conditions associated with (46) are equations (24b) and (27b) with $n = 2$. The solution of the problem is

$$Q^{(\alpha)}_{3(2)} = 12[R^{(\alpha)}g^{(\alpha)}(x_4)/n^{(\alpha)^2} + \delta_{\alpha 2}R^{(1)}/8] . \tag{47}$$

With the aid of (47) and the first order heat flux profile $Q^{(\alpha)}_{\iota(1)}$, obtained from (37) and (38), the heat flux microstructure is found to be

$$Q^{(\alpha)}_3 = Q_{3(0)} + \varepsilon P \cdot g^{(\alpha)}(x_4) + 12\varepsilon^2[R^{(\alpha)}g^{(\alpha)}(x_4)/n^{(\alpha)^2} + \delta_{\alpha 2}R^{(1)}/8] + o(\varepsilon^2) . \tag{48}$$

If equation (48) is averaged and $Q_{3(0)}$ is eliminated subsequently, the following simple result is obtained:

$$(Q^{(2a)}_3 - Q^{(1a)}_3)/\varepsilon^2 = R^{(1)} + R^{(2)} = R . \tag{49}$$

Equation (49), with $Q^{(\alpha a)}_3$ related to the averaged temperature through (17a), is the constitutive relation for R. The foregoing analysis completes the formulations of the basic differential equations for the mixture theory.

7. Summary of Mixture Equations

The mixture equations for the laminated composites can now be summarized as follows:

(a) Conservation of Energy:

$$Q^{(\alpha p)}_{i,i} - (-1)^\alpha P = -\mu^{(\alpha p)}\dot{T}^{(\alpha a)} , \tag{14}$$

$$M_{i,i} + (Q^{(2a)}_3 - Q^{(1a)}_3)/\varepsilon^2 = -\mu_{(m)}\dot{S}/12; \tag{32}$$

(b) Conduction Law and Other Constitutive Relations:

$$Q^{(\alpha p)}_i = -k^{(\alpha p)}T^{(\alpha a)}_{,i} + (-1)^\alpha \delta_{i3}k^{(\alpha)}S, \tag{17a}$$

$$M_\iota = -k_{(m)}S_{,\iota}/12 , \quad i = 1, 2 , \quad M_3 = P/12 \tag{43 a, b}$$

$$P = 12k_{(n)}(T^{(1a)} - T^{(2a)})/\varepsilon^2 - k_{(n)}S_{,3} \tag{41}$$

where

$$Q^{(\alpha p)}_i = n^{(\alpha)}Q^{(\alpha a)} , \quad \mu^{(\alpha p)} = n^{(\alpha)}\mu^{(\alpha)} , \quad k^{(\alpha p)} = n^{(\alpha)}k^{(\alpha)} , \tag{15), (17b}$$

$$\mu_{(m)} = \mu^{(1p)} + \mu^{(2p)} , \tag{33c}$$

$$k_{(n)} = 1/(n^{(1)}/k^{(1)} + n^{(2)}/k^{(2)}) , \quad k_{(m)} = k^{(1p)} + k^{(2p)} . \tag{42c), (43c}$$

If the constitutive relations are used to eliminate $Q^{(\alpha a)}_i$ and M_ι, the above set can be reduced to three differential equations with constant coefficients in the three unknows $T^{(1a)}$, $T^{(2a)}$ and S:

$$-k^{(1p)}T^{(1a)}_{,ii} - (k^{(1)} + k_{(n)})\, S_{,3} + 12k_{(n)}(T^{(1a)} - T^{(2a)})/\varepsilon^2 = -\mu^{(1p)}\dot{T}^{(1a)} , \tag{50}$$

$$-k^{(2p)}T^{(2a)}_{,ii} + (k^{(2)} + k_{(n)})\, S_{,3} - 12k_{(n)}(T^{(1a)} - T^{(2a)})/\varepsilon^2 = -\mu^{(2p)}\dot{T}^{(2a)} , \tag{51}$$

$$\begin{aligned}(k^{(1)}/n^{(1)} + k^{(2)}/n^{(2)})\, S + (k^{(1)} + k_{(n)})\, T^{(1a)}_{,3} - \\ -(k^{(2)} + k_{(n)})\, T^{(2a)}_{,3} - \varepsilon^2[k_{(m)}(S_{,11} + S_{,22}) + k_{(n)}S_{,33}]/12 = -\varepsilon^2\mu_{(m)}\dot{S}/12 .\end{aligned} \tag{52}$$

All coefficients in these equations are determined once the thermal properties of the constituents and their volume fractions are known.

From the form of the mixture equations a choice of dimensional quantities $\overline{\mu}_{(m)}$, $\overline{k}_{(m)}$, which are introduced in (4) to render nondimensional equations, can be made as:

$$\overline{\mu}_{(m)} = n^{(1)}\overline{\mu}^{(1)} + n^{(2)}\overline{\mu}^{(2)}; \quad \overline{k}_{(m)} = n^{(1)}\overline{k}^{(1)} + n^{(2)}\overline{k}^{(2)} . \tag{53}$$

8. Variational Principle and Boundary Conditions

Appropriate set of initial and boundary conditions has to be prescribed in order to apply the mixture equations to practical problems of the laminated composites. Due to the asymptotic nature of the mixture equations this question can be completely resolved only by a boundary layer analysis wherein the edge-zone solutions are matched with the mixture solutions. However, the need for a boundary layer analysis can be mitigated to some extent by use of a variational approach to determine the appropriate boundary data for the mixture differential equations; this is a common practice in plate and shell theories in elasticity.

The question of appropriate initial data to be used with the mixture theory can be resolved without any confusion, since only the first derivative with respect to time appears in the mixture equations as in the theories for homogeneous media. Thus, the appropriate initial data are

$$T^{(\alpha a)},\ S \quad \text{at} \quad t = 0 \quad \text{in } D\,. \tag{54}$$

Remaining set of appropriate boundary data is obtained by constructing a functional for the mixture equations. For simplicity, only steady state problems are considered. The functional is

$$\begin{aligned} \Pi = \iiint_D \Bigg[& \sum_{\alpha=1}^{2} \sum_{i=1}^{3} k^{(\alpha p)} \big(T^{(\alpha p)}_{,i} T^{(\alpha p)}_{,i}\big)/2 + 6k_{(n)}(T^{(1a)} - T^{(2a)})/\varepsilon^2 + (k^{(1)}/n^{(1)} + k^{(2)}/n^{(2)})\, S^2/2 + \\ & + \varepsilon^2 \left(k_{(m)} \sum_{\alpha=1}^{2} S_{,i} S_{,i} + k_{(n)} S_{,3} S_{,3}\right)/24 + S\{(k^{(1)} + k_{(n)})\, T^{(1a)}_{,3} - (k^{(2)} + k_{(n)})\, T^{(2a)}_{,3}\}\Bigg] \mathrm{d}x_1\, \mathrm{d}x_2\, \mathrm{d}x_3\,. \end{aligned} \tag{55}$$

If the first variation of $\delta\Pi$ for arbitrary $T^{(\alpha a)}$ and S is taken, the basic equations of the mixture theory (50—52), are obtained as the EULER equations:

$$\begin{aligned} \delta\Pi = & \iint_{\partial D} \left[\sum_{\alpha=1}^{2} \{k^{(\alpha p)} T^{(\alpha a)}_{,i} \nu_i - (-1)^\alpha\, (k^{(\alpha)} + k_{(n)})\, S\nu_3\}\, \delta T^{(\alpha a)} + \varepsilon^2 \{k_{(m)}(S_{,1}\nu_1 + S_{,2}\nu_2) + k_{(n)} S_{,3}\nu_3\}\, \delta S/12\right] \mathrm{d}A + \\ & + \iiint_D \Bigg[\sum_{\alpha=1}^{2} \{-(k^{(\alpha p)} T^{(\alpha a)}_{,i})_{,i} + (-1)^\alpha\, (k^{(\alpha)} + k_{(n)})\, S_{,3} - (-1)^\alpha\, 12k_{(n)}(T^{(1a)} - T^{(2a)})/\varepsilon^2\}\, \delta T^{(\alpha a)} + \\ & + \{(k^{(1)}/n^{(1)} + k^{(2)}/n^{(2)})\, S - \varepsilon^2(k_{(m)} \sum_{i=1}^{2} S_{,ii} + k_{(n)} S_{,33})/12 - \sum_{\alpha=1}^{2} (-1)^\alpha\, (k^{(\alpha)} + k_{(n)})\, T^{(\alpha a)}_{,3}\}\, \delta S\Bigg] \\ & \mathrm{d}x_1\, \mathrm{d}x_2\, \mathrm{d}x_3 = 0\,, \end{aligned} \tag{56}$$

where ν_i is an outward normal to the domain D on the boundary ∂D and $\mathrm{d}A$ is an infinitesimal surface element on ∂D.

Equation (56) also gives the appropriate set of boundary conditions to be prescribed on ∂D as follows:

$$\begin{aligned} & T^{(\alpha a)} \quad \text{or} \quad Q_1^{(\alpha p)} \nu_1 + Q_2^{(\alpha p)} \nu_2 + (Q_3^{(\alpha p)} + (-1)^\alpha\, k_{(n)} S)\, \nu_3\,, \\ & S \quad \text{or} \quad \varepsilon^2 \{k_{(m)}(S_{,1}\nu_1 + S_{,2}\nu_2) + k_{(n)} S_{,3}\nu_3\}/12\,. \end{aligned} \tag{57}$$

9. Recovery of Microstructure

Once initial boundary value problems defined by (50—52), (54), and (57) are solved to find $T^{(\alpha a)}$ and S, one can recover the temperature microstructure. To this end, one first calculates $T_{(0)}$, $Q_{3(0)}$ and $R^{(\alpha)}$.

From the averaged equation (40) one obtains $T_{(0)}$:

$$T_{(0)} = T^{(\alpha a)} - \varepsilon^2 (P v^{(\alpha a)} + S_{,3} w^{(\alpha a)})\,. \tag{58}$$

Similarly, $R^{(\alpha)}$ and $Q_{3(0)}$ are obtained from the $g^{(\alpha)}$-moment of (22):

$$M^{(\alpha)}_{i,i} + R^{(\alpha)} = -\mu^{(\alpha p)} \dot{S}/12\,, \tag{59}$$

and equation (43), and the averaged equation (48).

Finally, the field $T^{(\alpha)}$, $Q_3^{(\alpha)}$ correct to $O(\varepsilon^2)$ can be obtained from (40) and (48).

10. Decay of Spacially Harmonic Temperature Fields

In order to test the efficacy of the mixture model the decay of spacially harmonic temperature profiles is considered. The results obtained from the mixture equations (50—52) are compared with the exact ones.

Consider spacially harmonic temperature field of the form

$$\begin{Bmatrix} T^{(1a)} \\ T^{(2a)} \\ S \end{Bmatrix} = \begin{Bmatrix} \tilde{T}^{(1)} \\ \tilde{T}^{(2)} \\ iq\tilde{s} \end{Bmatrix} \exp\, \{iq(x_1 \cos \Theta + x_3 \sin \Theta) - \lambda t\} \tag{60}$$

in x_1, x_2-plane for the case where $(\ )_{,2} = 0$.

Substitution of (60) into the mixture equations (50—52) yields an eigenvalue problem for λ of the form

$$[L]\, U = \lambda U \tag{61}$$

where $U = (\tilde{T}^{(1)}, \tilde{T}^{(2)}, \tilde{s})^T$, and $[L]$ is a 3×3 matrix.

Exact decay relation for the harmonic temperature fields is obtained from equation (15) of reference [1]:

$$\cos(\varepsilon q \sin \Theta) - \cos(\varepsilon q n^{(1)} \varphi^{(1)}) \cos(\varepsilon q n^{(2)} \varphi^{(2)}) + \\ + \frac{1}{2} (\varepsilon q)^2 \left(\frac{k^{(2)}}{k^{(1)}} \varphi^{(1)^2} + \frac{k^{(1)}}{k^{(2)}} \varphi^{(2)^2} \right) \frac{\sin(\varepsilon q n^{(1)} \varphi^{(1)})}{\varepsilon q n^{(1)} \varphi^{(1)}} \frac{\sin(\varepsilon q n^{(2)} \varphi^{(2)})}{\varepsilon q n^{(2)} \varphi^{(2)}} = 0 \tag{62}$$

where

$$\varphi^{(\alpha)} = \left(\frac{\lambda}{q^2} \Gamma_a - \cos^2 \Theta \right)^{\frac{1}{2}}, \qquad \Gamma_{(\alpha)} = \mu^{(\alpha)} k_{(m)} / (k^{(\alpha)} \mu_{(m)}) . \tag{63}$$

The decay factor λ/q^2 has been calculated using (61) and (62) for the case where

$$n^{(2)}/n^{(1)} = 3 , \qquad \mu^{(2)}/\mu^{(1)} = 1/3 \tag{64}$$

and the results are shown in Figs. 3a and 3b.

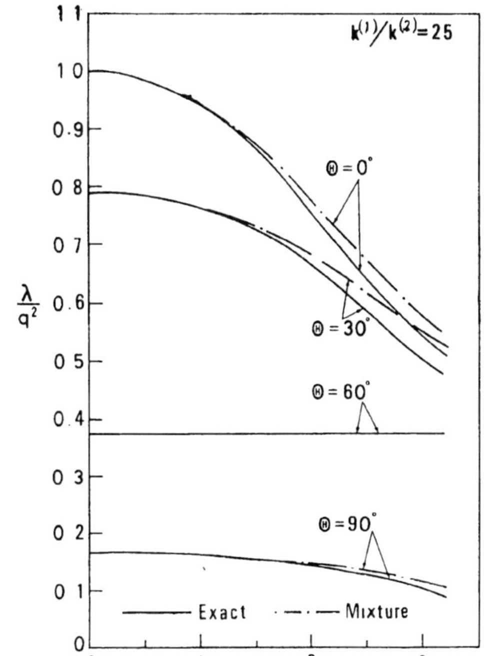

Fig. 3a. Time Decay Factors of Spacially Harmonic Temperature Fields with Different Wave Number for $\bar{k}^{(1)}/\bar{k}^{(2)} = 10$

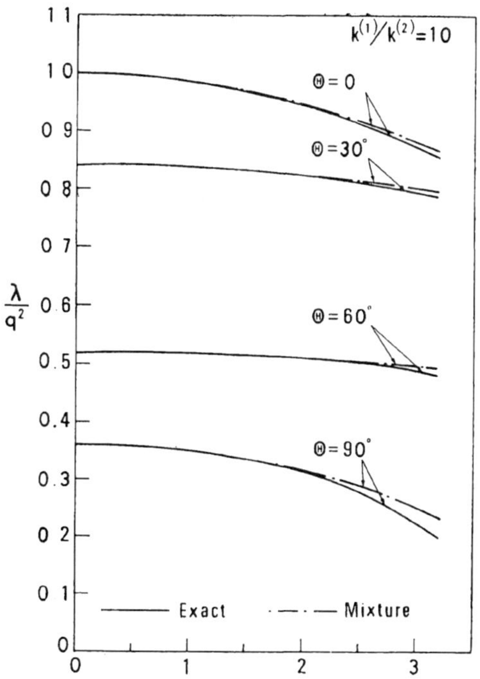

Fig. 3b. Time Decay Factors of Spacially Harmonic Temperature Fields with Different Wave Number for $\bar{k}^{(1)}/\bar{k}^{(2)} = 25$

It can be concluded that the decay factor λ/q^2 of the first mode is well approximated by the mixture theory down to wave-lengths equal to twice the composite cell length, i.e., $\varepsilon q \approx \pi$. This conclusion infers that the mixture equations can be used for unsteady state heat conduction problems without significant loss of accuracy.

10. Application of Mixture Equations to a Heat Shield Problem

Consider a half space of laminated composite with a lay-up angle φ as shown in Fig. 4. The coordinate system x, z selected in Fig. 4 is defined by the following transformation from x_1, x_3 to x, z:

$$x = x_1 \cos \varphi - x_3 \sin \varphi , \qquad z = x_1 \sin \varphi + x_3 \cos \varphi . \tag{65a), (65b}$$

An illustrative example treated here is a quiescent half space $x > 0$, subject to the boundary condition

$$\bar{T}^{(1a)} = \bar{T}^{(2a)} = 2\bar{T}_0 f(\bar{t}) , \qquad \bar{S} = 0 \quad \text{at} \quad x = 0 \tag{66a}$$

where

$$f(\bar{t}) = \begin{cases} 1, & 0 \leqq \bar{t} \leqq \bar{t}_0 \\ 0, & \bar{t}_0 < \bar{t} . \end{cases} \tag{66b}$$

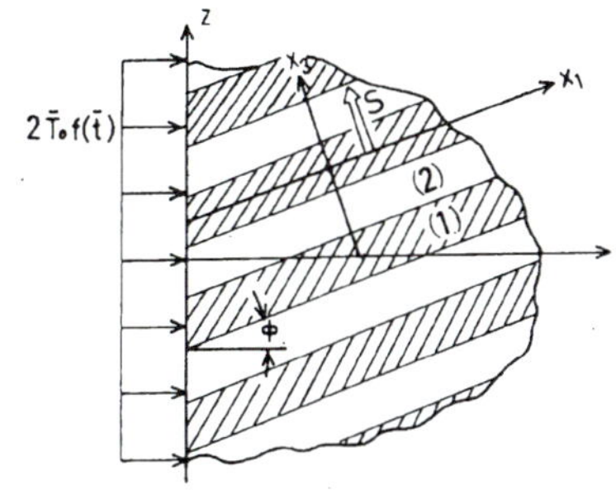

Fig. 4. Obliquely Laminated Half Space

Since there is no intrinsic length scale $\bar{\Lambda}$ in the half space problem, the pulse duration $\bar{t}_0$ $(= \bar{t}_{(m)})$ has been used to define the macrolength $\bar{\Lambda}$ for scaling in (4). Therefore, small values of ε correspond to short pulse length $\bar{t}_0$.

For this quasi-one-dimensional case the mixture equations obtained from the ones in Section 6 after coordinate transformation (65) are simplified as follows:

$$\frac{\partial}{\partial x} Q_x^{(\alpha p)} - (-1)^\alpha P = -\mu^{(\alpha p)} \dot{T}^{(\alpha a)} , \tag{67}$$

$$-\left(k_{(m)} \cos^2 \varphi \frac{\partial^2 S}{\partial x^2} + \sin \varphi \frac{\partial P}{\partial x}\right)\Big/12 + \left\{-\sin \varphi \, (Q_x^{(2a)} - Q_x^{(1a)}) + \cos \varphi (Q_z^{(2a)} - Q_z^{(1a)})\right\}\Big/\varepsilon^2 = -\mu_{(m)} \dot{S}/12 , \tag{68}$$

$$Q_x^{(\alpha p)} = n^{(\alpha)} Q_x^{(\alpha a)} = -k^{(\alpha)} \left(n^{(\alpha)} \frac{\partial}{\partial x} T^{(\alpha a)} + (-1)^\alpha S \sin \varphi\right), \tag{69a}$$

$$Q_z^{(\alpha p)} = n^{(\alpha)} Q_z^{(\alpha a)} = (-1)^\alpha k^{(\alpha)} S \cos \varphi , \tag{69b}$$

$$P = 12 k_{(n)} (T^{(1a)} - T^{(2a)})/\varepsilon^2 + k_{(n)} \frac{\partial S}{\partial x} \sin \varphi . \tag{70}$$

If the lay-up angle φ is set to be zero, equations (67) and (70) reduce to the mixture equations for heat flow in the direction parallel to the layering obtained by NAYFEH [3].

The boundary value problem defined above has been solved by using a finite difference scheme for the mixture equations (67—70). The material properties used for computations are given in Table 1. Calculations have been carried out for $\varepsilon^2 = 0.3$. The average temperature profiles in the two constituents, a short time before and after the termination of boundary heating, are shown in Figs. 5a and 5b. At this point, it is interesting to note that, in (60), λ/q^2 tends to the effective thermal diffusivity as $\varepsilon q \to 0$. Thus, from Figs. 3a, 3b it is found that the effective thermal diffusivity decreases as φ ranges from 0 to $\pi/2$. Consistently, the results in Figs. 5a, 5b show that

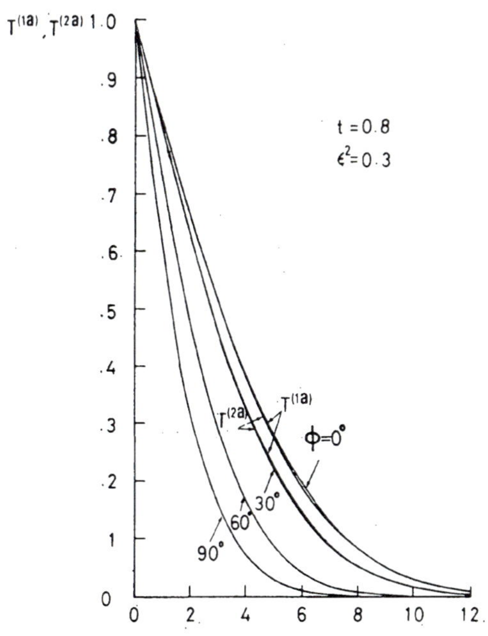

Fig. 5a. Profiles of Average Temperatures at $t = 0.8$ for Different Lay-Up Angles

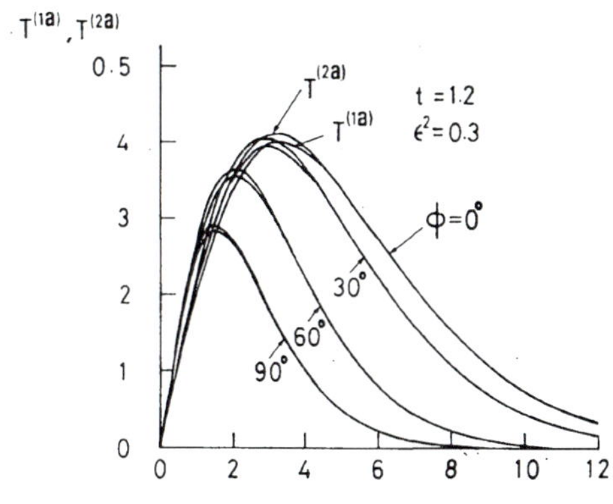

Fig. 5b. Profiles of Average Temperatures at $t = 1.2$ for Different Lay-Up Angles

thermal penetration depth decreases as φ ranges from 0 to $\pi/2$. Furthermore, the difference between the averaged temperatures is observed at a short time after the termination of boundary heating and becomes important as the lay-up angle φ decreases to zero. This kind of difference between the averaged temperatures has been confirmed in heat conduction of fiber reinforced composites [10], [11].

12. Concluding Remarks

The asymptotic method of multiple scales has been used to construct a continuum theory for heat conduction of periodically laminated media consisting of two constituents. The resulting theory was cast in the form of a binary mixture theory.

Comparison of exact versus mixture decay factors of spacially harmonic temperature fields with different wave number implies that the theory appears to be an attractive alternative to direct numerical solutions of unsteady state heat conduction in laminated composites where the energy is partitioned primarily into the first mode, and where material microdimension does not exceed one-half the dominant wavelength of the temperature profile.

Finally, the relation between this paper and reference [1] is clarified. A continuum model in [1] can achieve to model the dispersion phenomena in laminated composites to a desired accuracy keeping as many terms as necessary in the truncation process. The mixture theory can model the dispersion phenomena as well as the second order continuum model of reference [1]. However, the field equations of the continuum model in [1] include the spacial and temporal derivatives of the temperature which are higher than that occurs in the mixture theory. Consequently, questions about the specification of appropriate initial and boundary data are more difficult to solve than they are in the mixture formulation. An additional advantage of the present approach is that the derived field equations can be interpreted in a rather convenient fashion since it resembles to the field equations for homogeneous media except for the existance of interaction terms. Thus the choice of the two models should be made on a case by case basis.

Table 1. Material Properties Used for Computations

Thermal conductivity ratio:	$\bar{k}^{(1)}/\bar{k}^{(2)} = 50$
Heat capacity ratio:	$\bar{\mu}^{(1)}/\bar{\mu}^{(2)} = 0.5$
Volume fractions:	$n^{(1)} = n^{(2)} = 0.5$

References

1 MAEWAL, A., BACHE, T. C., and HEGEMIER, G. A., A Continuum Model for Diffusion in Laminated Composite Media, Journal of Heat Transfer, **100** (1978), S. 133–138.

2 HORVAY, G., MANI, R., VELUSWANI, M. A., and ZIENMEISTER, G. E., Transient Heat Conduction in Laminated Composites, Journal ofHeat Transfer, **95** (1973), S. 309–316.

3 NAYFEH, A. H., A Continuum Mixture Theory of Heat Conduction in Laminated Composites, Journal of Applied Mechanics, **42** (1975), S. 399–404.

4 NAYFEH, A. H., Continuum Modeling of Low Frequency Heat Conduction in Laminated Composites with Bonds, Journal of Heat Transfer, **102** (1980), S. 312–318.

5 HEGEMIER, G.A., GURTMAN, G. A., and NAYFEH, A. H., A Continuum Theory of Wave Propagation in Laminated and Fiber Reinforced Composites, International Journal of Solids and Structures, **9** (1973), S. 395–414.

6 HEGEMIER, G. A., MURAKAMI, H., and MAEWAL, A., On Construction of Mixture Theories for Composite Materials by the Method of Multi-Variable Asymptotic Expansions, Proceedings of Third International Symposium on Continuum Models of Discrete Systems, Freudenstadt, Germany, June 24–30, 1979.

7 MURAKAMI, H., MAEWAL, A., and HEGEMIER, G. A., A Mixture Theory with a Director for Linear Elastodynamics of Periodically Laminated Media, International Journal of Solids and Structures, **17** (1981), S. 155–173.

8 BRILLOUIN, L., Wave Propagation in Periodic Structures, Dover, 1953.

9 HEGEMIER, G. A., Mixture Theories with Microstructure for Wave Propagation and Diffusion in Composite Materials, Proceedings of Second International Symposium on Continuum Models of Discrete Systems, J. PROVAN (Editor), Montreal, 1977, S. 145–184.

10 MAEWAL, A., GURTMAN, G. A., and HEGEMIER, G. A., A Mixture Theory for Quasi-One-Dimensional Diffusion in Fiber Reinforced Composites, Journal of Heat Transfer, **100** (1978), S. 128–133.

11 MURAKAMI, H., HEGEMIER, G. A., and GURTMAN, G. A., A Nonlinear Mixture Theory for Quasi-One-Dimensional Heat Conduction in Fiber Reinforced Composites, International Journal of Solids and Structures, **16** (1980), S. 421–432.

Eingereicht am 5. 9. 1980

Anschrift: HIDENORI MURAKAMI, Associate Prof. of Applied Mechanics, Department of Mechanical Engineering, National Defense Academy, Yokosuka, Kanagawa, Japan;
AKHILESH MAEWAL, Assistant Professor, Department of Engineering and Applied Science, Yale University, New Haven, CT 06520, U.S.A.;
GILBERT A. HEGEMIER, Professor of Applied Mechanics, Department of Applied Mechanics and Engineering Sciences, University of California, San Diego, La Jolla, CA 92093, U.S.A.

ZAMM 61, 315 – 324 (1981)

J. ABOUDI / Y. BENVENISTE

An Average Theory for the Dynamic Behavior of a Laminated Elastic-Viscoplastic Work-Hardening Medium

Eine Mittelungstheorie wird entwickelt, die für besondere Belastungsarten das dynamische Verhalten eines elastisch-viskoplastischen geschichteten Mediums aus zwei Materialien unter Berücksichtigung der Verfestigung beschreibt. Jedes der beiden Materialien wird durch eine einheitliche Theorie der Elasto-Viskoplastizität mit Verfestigung festgelegt, welche weder ein Fließkriterium noch Belastungs-Entlastungsbedingungen benötigt. Die resultierende effektive Theorie ergibt ein System von nichtlinearen Differentialgleichungen für die gemittelten Spannungen, Verschiebungen und die geleistete plastische Arbeit. Die Theorie wird angewendet, um die effektiven Spannungs-Dehnungslinien des geschichteten Mediums zu berechnen, in denen der Übergang vom elastischen in den plastischen Bereich bestimmbar wird. Eine weitere Anwendung der Theorie liefert die dynamische Lösung eines geschichteten Stabes unter Impuls- oder Spannungsbelastung.

An average theory which models the dynamic behavior of a bilaminated medium under specific types of loading, made of elastic viscoplastic work-hardening materials is developed. Each constituent is represented by a unified theory of elasto-viscoplasticity including work-hardening, which requires neither a yield criterion nor loading or unloading conditions. The resulting effective theory appears in the form of a system of nonlinear differential equations for the avergage stresses, displacement and plastic work. The theory is applied to construct the effective stress-strain curves of the laminated medium in which the transition from the elastic to the plastic domain can be determined. The theory is also applied to obtain the dynamic response of a laminated slab subjected to velocity or stress input.

Развита теория усреднения, которая моделирует динамическое поведение двухслойной среды при специфическом типе нагружения и упруговязкопластическом упрочнении материала. Каждый компонент представлен с помощью унифицированной теории упруго-вязкопластичности включая упрочнение, которая не требует ни условий текучести, ни условий нагружения или разружения. Полученная эффективно теория выражается в виде системы нелинейных дифференциальных уравнений для средних напряжений, перемещений и пластической работы. Теория использована для построения кривых напряжения-деформации слоистой среды, по которым можно определить переход между упругой и пластической областями. Теория использована для вычисления динамической реакции слоистой пластины относительно скорости напряжения на входе.

1. Introduction

Equivalent models for the behavior of multi-phase media made of linearly elastic constituents are well established in the literature. These models replace the composite material by an equivalent homogeneous medium having effective elastic moduli and density, see for example HASHIN [1] and references cited there. For the special case of a periodically bilaminated material, the effective moduli have been given by POSTMA [2] and RYTOV [3] based on static and dynamic considerations, respectively.

Equivalent models for the behavior of laminated media made of elastic-plastic constituents are rather scarce, and some works deal specifically with the determination of the equivalent strength of the multi-phase medium [1]. For laminated and fiber reinforced elastic-perfectly plastic media, one-dimensional binary mixture models have been developed in [4] and [5] in the framework of the classical PRANDTL-REUSS equations.

In the present paper we propose an average theory for the dynamic behavior of a laminated medium made of elastic-viscoplastic work hardening constituents, under some specific types of loading. The elastic-viscoplastic behavior of the constituents is described by the unified theory of BODNER and PARTOM [6]. This theory represents the elastic-viscoplastic material by a single set of constitutive equations without a yield criterion, loading or unloading conditions. This strain rate dependent theory includes isotropic hardening. In the basic formulation, the viscoplastic material is characterized by five constants, the determination of which was summarized recently by MERZER and BODNER [7].

The laminated medium is subjected to a uniform normal loading in the direction of the lamination, and an asymptotic method is used to develop the average theory which represents the bilaminated medium. The effective theory is expressed in terms of a system of non-linear differential equations for the average stress components, average displacement and plastic work. The resulting model appears, obviously, also in a form of a unified theory which contains neither a yield criterion nor loading or unloading conditions. When the plasticity effects are completely neglected, the developed effective theory is shown to reduce to the equivalent modulus theory for a laminated medium made of perfectly elastic constituents thus forming a necessary check to the proposed model.

The method of solution of the resulting effective system of equations is numerical and is based on a finite difference procedure which provides the average displacement, stresses and plastic work.

The effective stress-strain curves of the equivalent model for different strain-rates are constructed, and the transition from the elastic region to the plastic region of the average stresses in the laminated medium is obtained.

The theory is applied to obtain the dynamic response of a laminated slab subjected to a uniform dynamic velocity or stress input at one of its surfaces while the other surface is kept rigidly clamped. The effect of the viscoplastic mechanism is studied by comparing the resulting field with that created in a laminated slab made of perfectly elastic constituents, and reflects itself essentially in the spreading and attenuation of the propagating pulse.

Basic Equations

Consider a periodic array of two alternating isotropic elastic-viscoplastic work hardening layers. Let $2h_1$ and $2h_2$ denote the width of each layer. The position of each particle in every layer is measured with respect to Cartesian coordinates $(x_1, x_2^{(\alpha)}, x_3)$ where $x_2^{(\alpha)}$ is measured from the mid-plane of layer α (see Fig. (1a)).

The constitutive equations of the elastic-viscoplastic work-hardening materials adopted in this paper are those proposed by Bodner and Partom [6] which have the important property that no yield criterion nor loading or unloading conditions are required. In this formulation both elastic and inelastic deformations are present at all stages of loading and unloading.

The constitutive equations of the material α ($\alpha = 1, 2$) can be described by separating the total strain rate components into elastic (reversible) and plastic (irreversible) strain rates as follows (Here and in the sequel the subscript or superscript (α) will indicate that quantities belong to either one of the constituents.)

$$\dot{\varepsilon}_{ij}^{(\alpha)} = \dot{\varepsilon}_{ij}^{(e\alpha)} + \dot{\varepsilon}_{ij}^{(p\alpha)}, \qquad i, j = 1, 2, 3 \tag{1}$$

where $\varepsilon_{ij}^{(\alpha)} = [u_{i,j}^{(\alpha)} + u_{j,i}^{(\alpha)}]/2$ with $u_i^{(\alpha)}$ being the components of the displacement vector (so that the strains are considered to be infinitesimal), dots represent time derivatives and $u_{i,j} = (\partial/\partial x_j)\, u_i$.

The elastic strain rates $\dot{\varepsilon}_{ij}^{(e\alpha)}$ are related to the stress rates $\dot{\sigma}_{ij}^{(\alpha)}$ according to the usual Hooke's law

$$\dot{\varepsilon}_{ij}^{(e\alpha)} = \dot{\sigma}_{ij}^{(\alpha)}/(2\mu_\alpha) - (\nu_\alpha/E_\alpha)\, \dot{\sigma}_{kk}^{(\alpha)}\delta_{ij} \tag{2}$$

where μ_α, ν_α, E_α are, respectively, the rigidity, the Poisson ratio and the Young modulus of the material, and δ_{ij} is the Kronecker delta.

The plastic strain rates $\dot{\varepsilon}_{ij}^{(p\alpha)}$ are related to the stresses according to the flow rule

$$\dot{\varepsilon}_{ij}^{(p\alpha)} = \dot{e}_{ij}^{(p\alpha)} = \Lambda_\alpha s_{ij}^{(\alpha)} \tag{3}$$

where $s_{ij}^{(\alpha)}$ and $\dot{e}_{ij}^{(p\alpha)}$ denote the deviators of the stress and plastic strain rate tensors respectively, i.e., $s_{ij}^{(\alpha)} = \sigma_{ij}^{(\alpha)} - \sigma_{kk}^{(\alpha)}\delta_{ij}/3$ and $\dot{e}_{ij}^{(p\alpha)} = \dot{\varepsilon}_{ij}^{(p\alpha)} - \dot{\varepsilon}_{kk}^{(p\alpha)}\delta_{ij}/3$. According to (3) the plastic deformations are necessarily incompressible, i.e., $\dot{\varepsilon}_{kk}^{(p\alpha)} = 0$.

Equation (3) can be squared to obtain Λ_α,

$$\Lambda_\alpha^2 = D_2^{(p\alpha)}/J_2^{(\alpha)} \tag{4}$$

where

$$D_2^{(p\alpha)} = \dot{e}_{ij}^{(p\alpha)}\dot{e}_{ij}^{(p\alpha)}/2, \qquad J_2^{(\alpha)} = s_{ij}^{(\alpha)}s_{ij}^{(\alpha)}/2 \tag{5}$$

which are the second invariants of the plastic strain rate deviator and the stress deviator tensors, respectively.

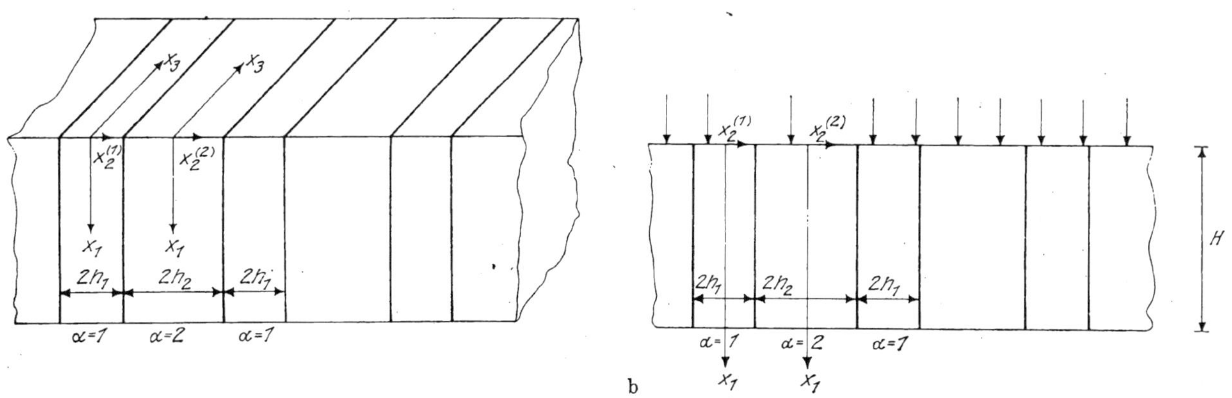

Fig. 1. (a) A laminated medium. (b) A laminated slab subjected to normal loads

Motivated by equations relating dislocation velocities and stresses, Bodner and Partom [6] proposed the relation

$$D_2^{(p\alpha)} = (D_0^{(\alpha)})^2 \exp\{-[z_\alpha^2/(3J_2^{(\alpha)})]^{n_\alpha} (n_\alpha + 1)/n_\alpha\} \tag{6}$$

where n_α is related to the steepness of the $D_2^{(p\alpha)} - J_2^{(\alpha)}$ curve, $(D_0^{(\alpha)})^2$ is the limiting value of $D_2^{(p\alpha)}$ for very high stresses and z_α is an internal state variable, referred to as the hardness of the material, which expresses its overall resistance to plastic flow. For isotropic work-hardening, the evolution equation for z_α is taken to depend on the amount of plastic (irreversible) work $w_p^{(\alpha)}$ which has been done on the material from a reference state. Specifically, z_α is assumed to have the form

$$z_\alpha = z_1^{(\alpha)} + (z_0^{(\alpha)} - z_1^{(\alpha)}) \exp[-m_\alpha w_p^{(\alpha)}/z_0^{(\alpha)}] \tag{7}$$

where $z_0^{(\alpha)}$, $z_1^{(\alpha)}$ and m_α are appropriate parameters of the material, and the rate of plastic work can be expressed in the form

$$\dot{w}_p^{(\alpha)} = \sigma_{ij}^{(\alpha)}\dot{\varepsilon}_{ij}^{(p\alpha)} = s_{ij}^{(\alpha)}\dot{\varepsilon}_{ij}^{(p\alpha)} = 2\Lambda_\alpha J_2^{(\alpha)}. \tag{8}$$

In (7), $z_0^{(\alpha)}$ is the initial hardness and $z_1^{(\alpha)}$ is the upper limit of z_α (saturation value) since otherwise, $D_2^{(p\alpha)}$ would approach zero for large $w_p^{(\alpha)}$ which leads to fully elastic behavior at appreciable strains.

Finally, the stresses fulfil, in the absence of body forces, the usual equations of motion

$$\varrho_\alpha \ddot{u}_i^{(\alpha)} = \sigma_{ij,j}^{(\alpha)} \tag{9}$$

where ϱ_α is the density of the material.

Development of the Average Theory

Let us now consider situations in which a quasi-dilatational type of wave motion in which the $u_1^{(\alpha)}$ displacement components are symmetric and the $u_2^{(\alpha)}$ displacement components are anti-symmetric with respect to the midplanes. Such deformations can be obtained, for example, when a laminated half-space is impacted by a uniform load normal to the layering in the x_1 direction (see Fig. (1b)).

Let us introduce the following nondimensionalization

$$\xi = x_1/l\,, \quad \eta^{(\alpha)} = x_2^{(\alpha)}/h\,, \quad U^{(\alpha)} = u_1^{(\alpha)}/l\,, \quad V^{(\alpha)} = u_2^{(\alpha)}/h\,, \tag{10}$$
$$\mathbf{S}^{(\alpha)} = \boldsymbol{\sigma}^{(\alpha)}/E_0\,, \quad W_p^{(\alpha)} = w_p^{(\alpha)}/E_0\,, \quad \tau = t/T$$

where

$$T = N_1/D_0^{(1)} + N_2/D_0^{(2)} \tag{11}$$

and

$$h = h_1 + h_2\,, \quad N_\alpha = h_\alpha/h\,, \quad E_0 = c_{11} + c_{22} + 2c_{12}$$

with

$$c_{\alpha\alpha} = N_\alpha(\lambda_\alpha + 2\mu_\alpha) - \lambda_\alpha^2/E' \quad \text{(no sum)}\,, \qquad c_{\alpha\beta} = \lambda_\alpha\lambda_\beta/E' \quad (\alpha, \beta = 1, 2, \alpha \neq \beta)\,,$$
$$E' = (\lambda_1 + 2\mu_1)/N_1 + (\lambda_2 + 2\mu_2)/N_2\,. \tag{12}$$

In fact E_0 can be considered to be simply a stress scale nondimensionalizing the stresses. It has been chosen in the above way so that

$$c_0 = (E_0/\varrho_0)^{1/2} \tag{13}$$

where $\varrho_0 = N_1\varrho_1 + N_2\varrho_2$, and c_0 is the so called representative effective velocity in a linearly elastic theory [2], [3].

The reference measure of length l in (10) is defined by $l = c_0 T$ where T is the reference time scale given by (11).

In the special case of a perfectly elastic laminated medium, the reference time scale T cannot be defined any more by (11) so instead we define there the reference length l as a characteristic length of the problem which is related to the time variation of the disturbance, and $T = l/c_0$.

Motivated by our previous work [8] on a perfectly elastic laminated medium, the nondimensional displacements and stresses are expanded in an asymptotic series in ε where $\varepsilon = h/l$ as follows

$$U^{(\alpha)} = \overset{(0)}{U}{}^{(\alpha)} + O(\varepsilon^2)\,, \qquad V^{(\alpha)} = \overset{(0)}{V}{}^{(\alpha)}\eta^{(\alpha)} + O(\varepsilon^2)\,, \tag{14), (15}$$

$$S_{11}^{(\alpha)} = \overset{(0)}{S}{}_{11}^{(\alpha)} + O(\varepsilon^2)\,, \quad S_{22}^{(\alpha)} = \overset{(0)}{S}{}_{22}^{(\alpha)} + O(\varepsilon^2)\,, \quad S_{33}^{(\alpha)} = \overset{(0)}{S}{}_{33}^{(\alpha)} + O(\varepsilon^2)\,, \quad S_{12}^{(\alpha)} = \overset{(0)}{S}{}_{12}^{(\alpha)}\eta^{(\alpha)}\varepsilon + O(\varepsilon^3)\,. \tag{16), (17), (18), (19}$$

Note that a superscript on the right-hand side denotes the specific consistuent ($\alpha = 1, 2$) and the one above denotes the order of the term in the asymptotic expansion. It is seen that the above expansions contain a $\eta^{(\alpha)}$ dependence reflecting the assumed symmetry properties of the displacement and the resulting stress components. It should be also noticed that in the plane-strain problems considered in this paper $S_{13}^{(\alpha)} = S_{23}^{(\alpha)} = 0$.

In the following we will construct an average theory modeling the dynamic deformation in the laminated medium by truncating the asymptotic series as appearing in (14—19) and producing results which are correct up to $O(\varepsilon^2)$ (i.e. not including terms of the order of ε^2). The resulting theory will predict the motion of the laminated medium as long as the characteristic length of the dynamic phenomenon is of the order of the characteristic length of the material l (defined by $l = c_0 T$). In the perfectly elastic case however an intrinsic material length does not exist, and the obtained average theory is valid when the ratio of h to a characteristic wave length l is very small.

We derive first a system of five quasilinear differential equations in the stress components. To this end we define the average nondimensional quantity

$$\bar{\chi}^{(\alpha)} = (1/N_\alpha) \int_0^{N_\alpha} \chi^{(\alpha)}\, d\eta^{(\alpha)}\,. \tag{20}$$

Then according to (14) and (16—18) we can write

$$\bar{U}^{(\alpha)} = \overset{(0)}{U}{}^{(\alpha)} + O(\varepsilon^2)\,, \qquad \bar{S}_{\gamma\gamma}^{(\alpha)} = \overset{(0)}{S}{}_{\gamma\gamma}^{(\alpha)} + O(\varepsilon^2) \quad \text{(no sum)}\,. \tag{21), (22}$$

The first two equations are obtained by employing the plane-strain condition $\varepsilon_{33}^{(\alpha)} = 0$ yielding from (1—3) in a nondimensional form

$$(E_0/2\mu_\alpha)\, S_{33,\tau}^{(\alpha)} - (\nu_\alpha E_0/E_\alpha)\, S_{,\tau}^{(\alpha)} + \Psi_\alpha(S_{33}^{(\alpha)} - S^{(\alpha)}/3) = 0 \tag{23}$$

where

$$S^{(\alpha)} = S_{11}^{(\alpha)} + S_{22}^{(\alpha)} + S_{33}^{(\alpha)} \tag{24}$$

and

$$\Psi_\alpha = E_0 T \Lambda_\alpha \,. \tag{25}$$

Introducing (21—22) in (23) we obtain up to $O(\varepsilon^2)$ the required first two equations ($\alpha = 1, 2$)

$$(E_0/2\mu_\alpha)\,\bar{S}^{(\alpha)}_{33,\tau} - (\nu_\alpha E_0/E_\alpha)\,\bar{S}^{(\alpha)}_{,\tau} + \overset{(0)}{\Psi}_\alpha(\bar{S}^{(\alpha)}_{33} - \bar{S}^{(\alpha)}/3) = 0 \tag{26}$$

with

$$\bar{S}^{(\alpha)} = \bar{S}^{(\alpha)}_{11} + \bar{S}^{(\alpha)}_{22} + \bar{S}^{(\alpha)}_{33} \,. \tag{27}$$

Furthermore, we have used in the above expansion the fact that $\Psi_\alpha = \Psi_\alpha(j_2^{(\alpha)})$ where $j_2^{(\alpha)} = J_2^{(\alpha)}/E_0^2 = \overset{(0)}{j_2^{(\alpha)}} + O(\varepsilon^2)$. It follows then that $\Psi_\alpha = \Psi_\alpha\Big(\overset{(0)}{j_2^{(\alpha)}}\Big) + O(\varepsilon^2)$ and $\overset{(0)}{\Psi}_\alpha = \Psi_\alpha\Big(\overset{(0)}{j_2^{(\alpha)}}\Big)$. Accordingly, using (4—7), $\overset{(0)}{\Psi}_\alpha$ is given by

$$\overset{(0)}{\Psi}_\alpha = D_0^{(\alpha)} T \exp\left\{-0.5\left[\overset{(0)}{H^2_\alpha}\Big/\Big(3\overset{(0)}{j_2^{(\alpha)}}\Big)\right]^{n_\alpha} (n_\alpha + 1)/n_\alpha\right\}\Big/\left[\overset{(0)}{j_2^{(\alpha)}}\right]^{1/2} \tag{28}$$

with

$$\overset{(0)}{H}_\alpha = \{z_1^{(\alpha)} + (z_0^{(\alpha)} - z_1^{(\alpha)}) \exp\,[-m_\alpha \overset{(0)}{W_p^{(\alpha)}}/(z_0^{(\alpha)}/E_0)]\}/E_0 \tag{29}$$

and

$$\overset{(0)}{j_2^{(\alpha)}} = [(\bar{S}^{(\alpha)}_{11} - \bar{S}^{(\alpha)}/3)^2 + (\bar{S}^{(\alpha)}_{22} - \bar{S}^{(\alpha)}/3)^2 + (\bar{S}^{(\alpha)}_{33} - \bar{S}^{(\alpha)}/3)^2]/2 \tag{30}$$

and an evolutionary equation for $\overset{(0)}{W_p^{(\alpha)}}$ will be given later.

The third equation is obtained from the continuity condition of the $V^{(\alpha)}$ displacements at the interface

$$N_1 \overset{(0)}{V^{(1)}} = -N_2 \overset{(0)}{V^{(2)}} \,. \tag{31}$$

Since

$$\varepsilon^{(\alpha)}_{22,\tau} = \frac{\partial^2 V^{(\alpha)}}{\partial \eta^{(\alpha)}\,\partial \tau} = \overset{(0)}{V^{(\alpha)}_{,\tau}} + O(\varepsilon^2) \tag{32}$$

using (15), we obtain

$$N_1 \varepsilon^{(1)}_{22,\tau} = -N_2 \varepsilon^{(2)}_{22,\tau} + O(\varepsilon^2) \,. \tag{33}$$

Applying (33) to (1—3) gives, using (22), in a nondimensional form

$$\begin{aligned}(E_0/2\mu_1)\,\bar{S}_{22,\tau} - (\nu_1 E_0/E_1)\,\bar{S}^{(1)}_{,\tau} + \overset{(0)}{\Psi}_1(\bar{S}_{22} - \bar{S}^{(1)}/3) = \\ = -(N_2/N_1)\,\{(E_0/2\mu_1)\,\bar{S}_{22,\tau} - (\nu_2 E_0/E_2)\,\bar{S}^{(2)}_{,\tau} + \overset{(0)}{\Psi}(\bar{S}_{22} - \bar{S}^{(\alpha)}/3)\}\end{aligned} \tag{34}$$

which is the required third equation, in which we have employed also the continuity condition of $S^{(\alpha)}_{22}$ at the interface

$$\overset{(0)}{S^{(1)}_{22}} = \overset{(0)}{S^{(2)}_{22}} + O(\varepsilon^2) \,, \tag{35}$$

which implies that

$$\bar{S}^{(1)}_{22} = \bar{S}^{(2)}_{22} + O(\varepsilon^2) \equiv \bar{S}_{22} \,. \tag{36}$$

The fourth differential equation is obtained by imposing the continuity condition for the $U^{(\alpha)}$ displacements at the interface, i.e.,

$$\overset{(0)}{U^{(1)}} = \overset{(0)}{U^{(2)}} + O(\varepsilon^2) \,, \tag{37}$$

which gives also

$$\bar{U}^{(1)} = \bar{U}^{(2)} + O(\varepsilon^2) \,. \tag{38}$$

Since,

$$\varepsilon^{(\alpha)}_{11} = \frac{\partial^2 U^{(\alpha)}}{\partial \xi\,\partial \tau} = \frac{\partial^2 \bar{U}^{(\alpha)}}{\partial \xi\,\partial \tau} + O(\varepsilon^2) \,, \tag{39}$$

we obtain

$$\frac{\partial^2 \bar{U}^{(1)}}{\partial \xi\,\partial \tau} = \frac{\partial^2 \bar{U}^{(2)}}{\partial \xi\,\partial \tau} + O(\varepsilon^2) \,. \tag{40}$$

Using this in the nondimensional forms of the constitutive equations (1—3) for both materials we get, after substituting the asymptotic expansions,

$$(E_0/2\mu_1)\,\bar{S}^{(1)}_{11,\tau} - (\nu_1 E_0/E_1)\,\bar{S}^{(1)}_{\tau,} + \overset{(0)}{\Psi}_1(\bar{S}^{(1)}_{11} - \bar{S}^{(1)}/3) = (E_0/2\mu_2)\,\bar{S}^{(2)}_{11,\tau} - (\nu_2 E_0/E_2)\,\bar{S}^{(2)}_{,\tau} + \overset{(0)}{\Psi}_2(\bar{S}^{(2)}_{11} - \bar{S}^{(2)}/3) \,. \tag{41}$$

The fifth differential equation is obtained by defining

$$\bar{U} = \bar{U}^{(1)} = \bar{U}^{(2)} \tag{42}$$

and using the constitutive equation for $\varepsilon_{11,\tau}$ (eqs. (1—3)) in a nondimensionalized form after employing the asymptotic expansions. We obtain

$$\frac{\partial^2}{\partial\xi\,\partial\tau}\overline{U} = (E_0/2\mu_1)\,\overline{S}^{(1)}_{11,\tau} - (\nu_1 E_0/E_1)\,\overline{S}^{(1)}_{,\tau} + (\overline{S}^{(1)}_{11} - \overline{S}^{(1)}/3)\,\overset{(0)}{\Psi}_1\,. \tag{43}$$

By using the relation

$$\overline{S}_{11} = N_1\overline{S}^{(1)}_{11} + N_2\overline{S}^{(2)}_{11} \tag{44}$$

in the above five equations (26), (34), (41), (43) we get five differential equations in the five average stresses $\overline{S}_{11}$, $\overline{S}^{(2)}_{11}$, $\overline{S}_{22}$, $\overline{S}^{(1)}_{33}$, $\overline{S}^{(2)}_{33}$. These equations can be written in the compact form

$$\boldsymbol{A}\boldsymbol{X}_{,\tau} = \boldsymbol{F} \tag{45}$$

where $\boldsymbol{X}$ is the vector of the above five unknowns, $\boldsymbol{A}$ is 5×5 matrix whose elements are given in the Appendix and $\boldsymbol{F}$ is a vector whose components are given in terms of the stresses, plastic work and $\overline{U}_{,\xi\tau}$ and are again given in the Appendix.

The plastic work is governed by the following system of two differential equations

$$\frac{\partial}{\partial\tau}\overset{(0)}{W}{}^{(\alpha)}_{p} = 2\overset{(0)}{\Psi}_{\alpha}\overset{(0)}{j}{}^{(\alpha)}_{2}\,. \tag{46}$$

These seven first order quasi-linear differential equations should be suplemented by the dynamic equation for the laminated medium. The substitution of the asymptotic series (14—19) in the nondimensional form of the first two equations of motion (9), give:

$$\frac{\partial}{\partial\xi}\overset{(0)}{S}{}^{(\alpha)}_{11} + \overset{(1)}{S}{}^{(\alpha)}_{12} = (\varrho_\alpha/\varrho_0)\,\overset{(0)}{U}{}^{(\alpha)}_{,\tau\tau}\,, \qquad \frac{\partial}{\partial\xi}\overset{(1)}{S}{}^{(\alpha)}_{12} + 2\overset{(2)}{S}{}^{(\alpha)}_{22} = (\varrho_\alpha/\varrho_0)\,\overset{(0)}{V}{}^{(\alpha)}_{,\tau\tau}\,. \tag{47), (48}$$

The second equation (48) is redundant in the framework of our average theory in which the asymptotic expansion is carried up to $O(\varepsilon^2)$, since it will contain higher order terms through $\overset{(2)}{S}{}^{(\alpha)}_{22}$. Accordingly, we obtain from (47) the two equations

$$\frac{\partial}{\partial\xi}\overline{S}^{(\alpha)}_{11} + \overset{(1)}{S}{}^{(\alpha)}_{12} = (\varrho_\alpha/\varrho_0)\,\overline{U}_{,\tau\tau}\,. \tag{49}$$

Multiplying the first equation in (49) (with $\alpha = 1$) by N_1 and the second one (with $\alpha = 2$) by N_2, summing the results and taking into account the fact that the shearing stresses are continuous accross the interfaces of the layers, i.e., $N_1\overset{(1)}{S}{}^{(1)}_{12} = -N_2\overset{(1)}{S}{}^{(2)}_{12}$, we obtain the required dynamic equation

$$\overline{U}_{,\tau\tau} = \frac{\partial}{\partial\xi}\overline{S}_{11} \tag{50}$$

in terms of the average displacement and stresses.

The dynamic equation (50) together with the above seven equations (45—46), and the appropriate initial and boundary conditions, forming altogether the average theory, describe completely the motion of the laminated medium in the effective sense.

For a laminated slab at rest for $t < 0$ whose thickness is H, which is subject to an extended loading on its surface $x_1 = 0$ while the other surface is held rigidly clamped (see Fig. (1b)), the following boundary conditions are considered in this paper

$$\text{(a)}\quad \overline{U}_{,\tau}(0,\tau) = f(\tau)\,, \qquad \overline{U}(H/l,\tau) = 0 \tag{51}$$

for a velocity input $f(\tau)$ at $\xi = 0$ ($f(\tau) = 0$ for $\tau \leqq 0$). If, on the other hand, we take the case of an applied stress on the surface $\xi = 0$, we have

$$\text{(b)}\quad \overline{S}_{11}(0,\tau) = g(\tau)\,, \qquad \overline{U}(H/l,\tau) = 0 \tag{52}$$

for a stress input $g(\tau)$ at $\xi = 0$ ($g(\tau) = 0$ for $\tau \leqq 0$).

Special Cases

The average theory which was developed in the previous section contains the following two special cases which in fact, constitute a necessary check to the theory.

(a) Perfectly elastic layers

When the materials forming the laminated medium are perfectly elastic, $D^{(\alpha)}_0 = 0$ and $\boldsymbol{F}$ in (45) (whose components are given in the viscoplastic case in the Appendix) is given this time by $F_1 = F_2 = F_3 = F_4 = 0$, $F_5 = \frac{\partial^2}{\partial\xi\,\partial\tau}\overline{U}$.

The matrix $\boldsymbol{A}$ in (45) is not affected in this special case.

It can be shown by inverting $\boldsymbol{A}$ that the following equation is obtained

$$\overline{S}_{11,\tau} = F_5 \tag{53}$$

which implies the relation

$$\overline{S}_{11} = \frac{\partial}{\partial \xi} \overline{U} \tag{54}$$

which is the constitutive equation in a nondimensional form in the framework of the equivalent modulus theory of a perfectly elastic laminated medium under our specific type of loading, see POSTMA [2].

(b) Single elastic-viscoplastic material

When the two materials are identical, the laminated medium reduces to a homogeneous elastic-viscoplastic work-hardening one. For the one-dimensional problems considered here the constitutive relations (1—3) give the following two independent equations in a nondimensional form

$$\begin{aligned} &E_0(1/2\mu - 2\nu/E)\, S_{33,\tau} - (\nu E_0/E)\, S_{11,\tau} = (S/3 - S_{33})\, \Psi \\ &-(2E_0\nu/E)\, S_{33,\tau} + E_0(1/2\mu - \nu/E)\, S_{11,\tau} = (S/3 - S_{11})\, \Psi + \frac{\partial^2}{\partial\xi\, \partial\tau} U \end{aligned} \tag{55}$$

where obviously the subscripts and the superscripts α have been omited, and S_{ij} and U are the stresses and displacement in the homogeneous medium.

These equations can be obtained directly from the average equations (45) by noting that in passing to the homogeneous medium we have used the fact that $\overline{S}_{11}^{(1)} = \overline{S}_{11}^{(2)} = \overline{S}_{11} = S_{11}$, $\overline{S}_{22} = S_{22}$, $\overline{S}_{33}^{(1)} = \overline{S}_{33}^{(2)} = S_{33}$, $\overline{U} = U$ and $\overset{(0)}{\Psi} = \Psi$. Consequently, the five equations in (45) yield directly the two equations (55). As to the other three equations for the plastic work and the dynamic equation (46) and (50) they reduce, obviously, to the appropriate equations of the homogeneous medium.

Numerical Treatment

The dynamic equation of motion (50) and the associated seven quasi-linear differential equations (45—46) for the stresses and plastic work are solved numerically using a finite difference procedure. To this end we divide the interval $0 \leqq \xi \leqq H/l$ into N equal segments of length $\Delta\xi$, and introduce a time increment $\Delta\tau$. The procedure can be divided into three parts as follows.

(1) The equation of motion is employed to determine the average displacement $\overline{U}$ at any internal point $0 < \xi < H/l$ by approximating the dynamic equation (50) by the finite difference equation

$$\overline{U}(\xi, \tau + \Delta\tau) = 2\overline{U}(\xi, \tau) - \overline{U}(\xi, \tau - \Delta\tau) + r^2[\overline{S}_{11}(\xi + \Delta\xi, \tau) - \overline{S}_{11}(\xi - \Delta\xi, \tau)] \tag{56}$$

with $r^2 = (\Delta\tau)^2/(2\Delta\xi)$.

According to (56) it it possible to compute $\overline{U}$ at the time $\tau + \Delta\tau$ at any internal point from the values of $\overline{U}$ at the times τ and $\tau - \Delta\tau$ and the values of $\overline{S}_{11}$ at τ. This equation approximates (50) with a truncation error of the second order in the time and spatial increments.

(2) The stresses and plastic work at the internal points at time $\tau + \Delta\tau$ are computed by using the improved EULER-CAUCHY method for the numerical solution of the system of the seven differential equations (45—46) as follows. First we use the predictor formulae which gives tentatively the values

$$\overset{*}{\boldsymbol{X}}(\xi, \tau + \Delta\tau) = \boldsymbol{X}(\xi, \tau) + \Delta\tau \boldsymbol{A}^{-1} \overset{*}{\boldsymbol{F}}(\xi, \tau)\,, \qquad \overset{(0)}{W}{}_p^{(\alpha)}(\xi, \tau + \Delta\tau) = \overset{(0)}{W}{}_p^{(\alpha)}(\xi, \tau) + 2\Delta\tau\, \overset{(0)}{\Psi}_\alpha(\xi, \tau)\, \overset{(0)}{j}{}_2^{(\alpha)}(\xi, \tau)\,. \tag{57}$$

These values are corrected by:

$$\begin{aligned} &\boldsymbol{X}(\xi, \tau + \Delta\tau) = \boldsymbol{X}(\xi, \tau) + \Delta\tau \boldsymbol{A}^{-1}\Big[\boldsymbol{F}(\xi, \tau) + \overset{*}{\boldsymbol{F}}(\xi, \tau + \Delta\tau)\Big]/2\,, \\ &\overset{(0)}{W}{}_p^{(\alpha)}(\xi, \tau + \Delta\tau) = \overset{(0)}{W}{}_p^{(\alpha)}(\xi, \tau) + \Delta\tau\Big[\overset{(0)}{\Psi}_\alpha(\xi, \tau)\, \overset{(0)}{j}{}_2^{(\alpha)}(\xi, \tau) + \overset{*}{\overset{(0)}{\Psi}}_\alpha(\xi, \tau + \Delta\tau)\, \overset{*}{\overset{(0)}{j}}{}_2^{(\alpha)}(\xi, \tau + \Delta\tau)\Big] \end{aligned} \tag{58}$$

where the stars on $\boldsymbol{F}$, $\overset{(0)}{\Psi}_\alpha$, $\overset{(0)}{j}{}_2^{(\alpha)}$ indicate that these quantities should be computed using the predicted values given by (57). These equations are of a second order accuracy in the time increment.

(3) The dependent variables at the boundary points $\xi = 0$ and $\xi = H/l$ are computed according to whether a velocity input (51) or stress input (52) are applied.

When the velocity input (51) is applied at $\xi = 0$, the average displacement at the end points are given directly by

$$\overline{U}(0, \tau + \Delta\tau) = \int_0^{\tau+\Delta\tau} f(\lambda)\, \mathrm{d}\lambda\,, \qquad \overline{U}(0, \tau + \Delta\tau) = 0 \tag{59}$$

and the stresses and plastic work can be computed at these end points by using the same procedure (57—58) for the internal points.

When the stress input (52), on the other hand, is applied at $\xi = 0$, the procedure for the computation of the stresses and the displacement at $\xi = 0$ should be modified since $\frac{\partial^2}{\partial\xi\,\partial\tau}\overline{U}$ in $\boldsymbol{F}$ of (45) is unknown, but $\overline{S}_{11}$ is known this time. Consequently, the system of equations (45) takes now the form

$$\boldsymbol{B}\boldsymbol{Y}_{,\tau} = \boldsymbol{G} \tag{60}$$

where $\boldsymbol{Y}$ is the vector of unknowns

$$\boldsymbol{Y} = \left\{\frac{\partial}{\partial\xi}\overline{U},\ \overline{S}_{11}^{(2)},\ \overline{S}_{22},\ \overline{S}_{33}^{(1)},\ \overline{S}_{33}^{(2)}\right\} \tag{61}$$

and

$$\boldsymbol{G} = \{F_1 - A_{11}g(\tau),\ F_2 - A_{21}g(\tau),\ F_3 - A_{31}g(\tau),\ F_4 - A_{41}g(\tau),\ (\overline{S}^{(1)}/3 - \overline{S}_{11}^{(1)})\overset{(0)}{\Psi_1} - A_{51}g(\tau)\}\,. \tag{62}$$

The matrix $\boldsymbol{B}$ in (60) is equal to $\boldsymbol{A}$ but with $B_{11} = B_{21} = B_{31} = B_{41} = 0$ and $B_{51} = -1$.

The system of seven differential equations (60) and (46) are solved by the improved Euler-Cauchy method in order to obtain $\overline{S}_{11}^{(2)}, \overline{S}_{22}, \overline{S}_{33}^{(1)}, \overline{S}_{33}^{(2)}, \overset{(0)}{W_p^{(\alpha)}}$ at the point $\xi = 0$ at time $\tau + \Delta\tau$. The displacement $\overline{U}(0, \tau + \Delta\tau)$ is determined from $\frac{\partial^2}{\partial\xi\,\partial\tau}\overline{U}$ which is the first unknown in the system (60).

Results

In this paper we give results for a laminated medium made of titanium ($\alpha = 1$) and copper ($\alpha = 2$). The appropriate parameters of these materials are given by [6], [9]:

$$\nu_1 = 0.34\,,\quad \mu_1 = 0.44 \times 10^{11}\ \mathrm{N/m^2}\,,\quad D_0^{(1)} = 10^4\ \mathrm{sec^{-1}}\,,\quad n_1 = 1\,,\quad z_0^{(1)} = 1.15 \times 10^9\ \mathrm{N/m^2}\,,$$
$$z_1^{(1)} = 1.4 \times 10^9\ \mathrm{N/m^2}\,,\quad m_1 = 100\,,\quad \varrho_1 = 4.87 \times 10^3\ \mathrm{kg/m^3};$$
$$\nu_2 = 0.33\,,\quad \mu_2 = 0.45 \times 10^{11}\ \mathrm{N/m^2}\,,\quad D_0^{(2)} = 10^4\ \mathrm{sec^{-1}}\,,\quad n_2 = 7.5\,,\quad z_0^{(2)} = 0.63 \times 10^8\ \mathrm{N/m^2}\,,$$
$$z_1^{(2)} = 2.5 \times 10^8\ \mathrm{N/m^2}\,,\quad m_2 = 8.19\,,\quad \varrho_2 = 8.96 \times 10^3\ \mathrm{kg/m^3}\,.$$

Note that the elastic parameters of these materials are almost identical, but there is a considerable contrast between the constants which describe their plastic behavior.

The reinforcement ratios are chosen to be $N_1 = N_2 = 1/2$. For these values we obtain from (12) $E_0 = 1{,}79 \times 10^{11}\ \mathrm{N/m^2}$ and $\varrho_0 = 6.91 \times 10^3\ \mathrm{kg/m^3}$ so that (13) gives $c_0 = 5100$ m/sec which is the effective wave speed of a deformation in the x_1 direction propagating in the laminae direction [3]. This value of c_0 implies that the intrinsic length of the viscoplastic laminated material is $l = 0.51$ m.

The width of the slab H is chosen to be equal to l, and the spatial increment is taken to be $\Delta\xi = 0.02$ for which a time increment $\Delta\tau = 0.001$ provided satisfactory accurate solution.

(1) Effective stress-strain curves

The system of quasi-linear differential equations (45—46) can be employed to construct the effective stress-strain curves for the laminated medium, i.e., plots of the average stress $\overline{S}_{11}$ versus the average strain $\overline{\varepsilon}_{11} = \frac{\partial}{\partial\xi}\overline{U}$ for a given value of the strain rate $\overline{\varepsilon}_{11,\tau}$. This is achieved by choosing $\overline{\varepsilon}_{11,\tau} = e_0$ (so that $\overline{\varepsilon}_{11} = e_0\tau$) and substituting it in the system (45—46). This system of seven differential equations is then solved numerically using, for example, the fourth order Runge-Kutta method which gives the average stresses $\boldsymbol{X}$ and the plastic work $\overset{(0)}{W_p^{(\alpha)}}$.

In Fig. (2) we present the effective stress-strain curve for a laminated medium made of titanium and copper. In the same Figure we give also for comparison the stress-strain curves for single homogeneous materials: titanium and copper (in these two cases $\overline{S}_{11} = S_{11}$ and $\overline{\varepsilon}_{11} = \varepsilon_{11}$). The strain rate for all these curves is taken to be $e_0 = 10^{-6}$ which corresponds to $10^{-2}\ \mathrm{sec^{-1}}$. All the stress values in the Figure are normalized with respect to E_0 ($E_0 = 1.79 \times 10^{11}\ \mathrm{N/m^2}$). The change of the slope from unity in the elastic range to the relevant values in the plastic range is clearly seen in the Figure in all three cases. The elastic range in the case of the titanium is seen to be much more extended than that of the copper, and the effective stress-strain curve of the laminated medium falls between these two curves. In all curves the point of yielding can be quite clearly determined. Therefore, the present theory provides, in particular, the effective yield point of the two-phase laminated medium.

(2) The dynamic response of the laminated slab

When the laminated slab is subjected to a velocity input at $x_1 = 0$ and is kept rigidly clamped at $x_1 = H$, the boundary conditions are given by (51).

For a loading input, we choose in (51) the function $f(\tau)$ in the form

$$f(\tau) = \begin{cases} f_0 \sin \pi\tau(/2\tau_m) & 0 \leqq \tau \leqq \tau_m \\ f_0 & \tau > \tau_m \end{cases} \tag{63}$$

with $f_0 = 0.01$ and $\tau_m = 0.5$.

In Fig. (3) the resulting average velocity $\bar{U}_{,\tau}$ is shown versus τ at the observation points $\xi = 0.3$ and $\xi = 0.7$ within the laminated slab. This is compared with the velocity field obtained in a laminated slab made of titanium and copper but assuming that their behavior is perfectly elastic. The basic effect of the viscoplastic mechanism is in the spreading and attenuation of the propagating pulse. It should be noticed that the dynamic responses exhibited in the Figure contain both the direct and the reflected wave.

In Fig. (4) we present the dynamic response of the laminated slab for the velocity input

$$f(\tau) = \begin{cases} f_0 \sin \pi\tau/\tau_m & 0 \leqq \tau \leqq \tau_m \\ 0 & \tau > \tau_m \end{cases} \tag{64}$$

with $f_0 = 0.01$ and $\tau_m = 1$.

In this case there is an unloading phenomena caused directly by the applied input. The present unified theory does not require a condition for unloading and it treats this phenomena in the same framework. The resulting effects of viscoplasticity are the same as those appearing in Fig. (3).

In Fig. (5) the average stress field $\bar{S}_{11}$ is shown versus τ at the same observation points when a stress input (52) is applied to the laminated slab. Here we choose $g(\tau)$ to be equal to $f(\tau)$ in (63). In this case the modification

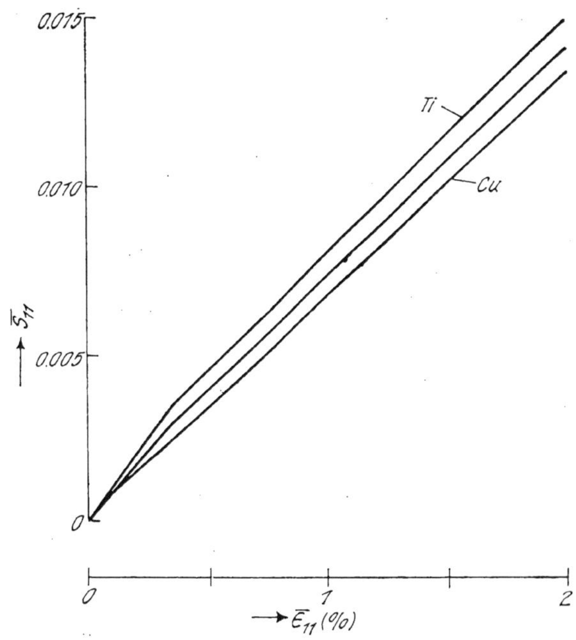

Fig. 2. Effective stress-strain curve for an elastic-viscoplastic laminated medium made of titanium and copper, and the stress-strain curves for a homogeneous medium made of (a) titanium (Ti), (b) copper (Cu). The strain rate in all cases is 10^{-2} sec^{-1}

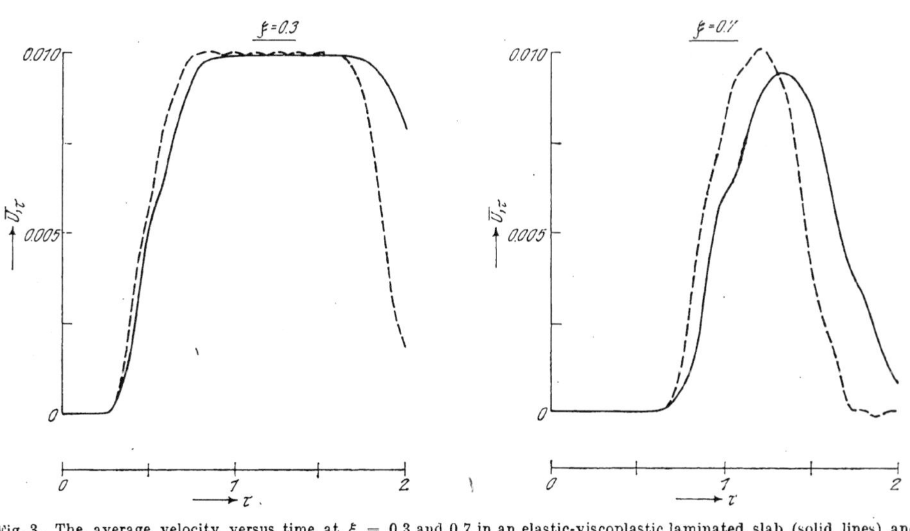

Fig. 3. The average velocity versus time at $\xi = 0.3$ and 0.7 in an elastic-viscoplastic laminated slab (solid lines) and in a perfectly elastic laminated slab (dashed lines). The velocity input is given by (63).

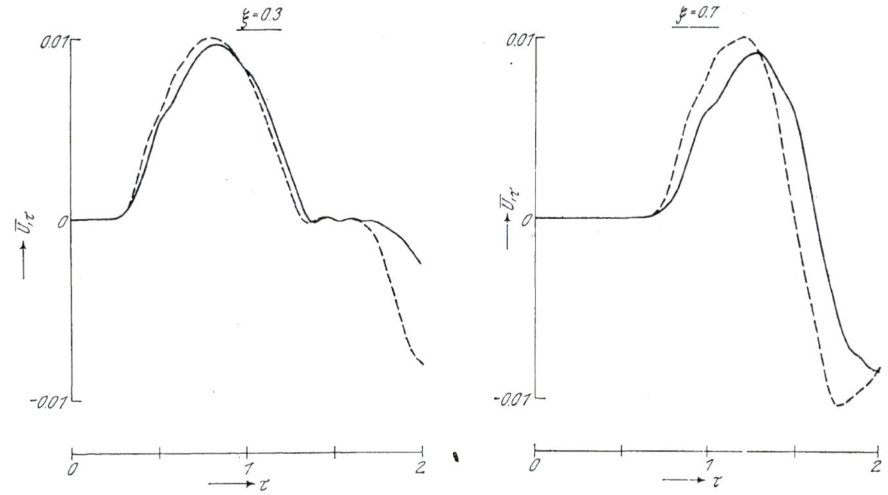

Fig. 4. Same as Fig. (3) but for the velocity input (64)

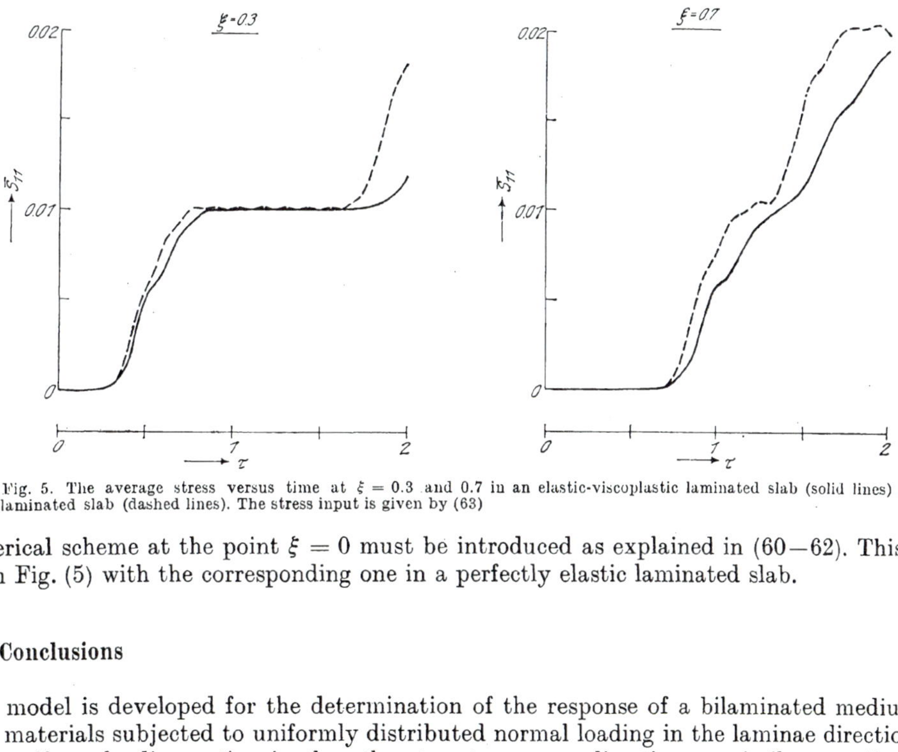

Fig. 5. The average stress versus time at $\xi = 0.3$ and 0.7 in an elastic-viscoplastic laminated slab (solid lines) and in a perfectly elastic laminated slab (dashed lines). The stress input is given by (63)

of the numerical scheme at the point $\xi = 0$ must be introduced as explained in (60–62). This stress field is also compared in Fig. (5) with the corresponding one in a perfectly elastic laminated slab.

Conclusions

An average model is developed for the determination of the response of a bilaminated medium made of elastic-viscoplastic materials subjected to uniformly distributed normal loading in the laminae direction.

For a uniform loading acting in the other two transverse directions, a similar one-dimensional system of equations can be developed which models the dynamic response of the composite medium. In this case the development of the theory is basically similar but more complicated.

Appendix

The elements of the 5×5 matrix $\boldsymbol{A}$ of (45) are given by

$$A_{11} = -\nu_1 E_0/(E_1 N_1)\,, \quad A_{12} = -N_2 A_{11}\,, \quad A_{13} = -\nu_1 E_0/E_1\,, \quad A_{14} = A_{13} + E_0/(2\mu_1)\,, \quad A_{15} = 0\,,$$
$$A_{21} = 0\,, \quad A_{22} = -\nu_2 E_0/E_2\,, \quad A_{23} = A_{22}\,, \quad A_{24} = 0\,, \quad A_{25} = A_{22} + E_0/(2\mu_2)\,,$$
$$A_{31} = A_{11}\,, \quad A_{32} = A_{12} + A_{35}\,, \quad A_{33} = A_{14} + N_2 E_0/(2N_1\mu_2) + A_{35}\,, \quad A_{34} = A_{13}\,, \quad A_{35} = -N_2 E_0 \nu_2/(N_1 E_2)\,,$$
$$A_{41} = E_0/(2\mu_1 N_1) + A_{11}\,, \quad A_{42} = -E_0 N_2/(2\mu_1 N_1) + A_{12} - E_0/(2\mu_2) - A_{23}\,, \quad A_{43} = A_{13} - A_{23}\,, \quad A_{44} = A_{13}\,,$$
$$A_{45} = -A_{22}\,, \quad A_{51} = A_{41}\,, \quad A_{52} = -N_2 A_{41}\,, \quad A_{53} = A_{13}\,, \quad A_{54} = A_{13}\,, \quad A_{55} = 0\,.$$

The components of the vector $\boldsymbol{F}$ in (45) are given by

$$F_1 = [\bar{S}^{(1)}/3 - \bar{S}_{33}^{(1)}]\,\overset{(0)}{\Psi}_1\,, \quad F_2 = [\bar{S}^{(2)}/3 - \bar{S}_{33}^{(2)}]\,\overset{(0)}{\Psi}_2\,, \quad F_3 = [\bar{S}^{(1)}/3 - \bar{S}_{22}]\,\overset{(0)}{\Psi}_1 - (N_2/N_1)\,\overset{(0)}{\Psi}_2[\bar{S}_{22} - \bar{S}^{(2)}/3]\,,$$
$$F_4 = [\bar{S}^{(1)}/3 - \bar{S}_{11}^{(1)}]\,\overset{(0)}{\Psi}_1 + [\bar{S}_{11}^{(2)} - \bar{S}^{(2)}/3]\,\overset{(0)}{\Psi}_2\,, \quad F_5 = \frac{\partial^2}{\partial\xi\,\partial\tau}\,\bar{U} - [\bar{S}_{11}^{(1)} - \bar{S}^{(1)}/3]\,\overset{(0)}{\Psi}_1$$

with $\bar{S}^{(\alpha)}$ and $\overset{(0)}{\Psi}_\alpha$ defined in (27) and (28) respectively.

References

1 Z. HASHIN, Theory of Fiber Reinforced Materials, NASA Report on Contract NASA 1–8818 (1970).
2 G. W. POSTMA, Wave Propagation in a Stratified Medium, Geophysics **20**, 780–806 (1955).
3 S. M. RYTOV, Acoustical Properties of a Thinly Laminated Medium, Soviet Phys., Acoust. **2**, 68–80 (1956).
4 G. A. HEGEMIER, Finite-Amplitude Elastic-Plastic Wave Propagation in Laminated Composites, J. Appl. Phys., **45**, 4248–4253 (1974).
5 G. A. HEGEMIER and GURTMAN, G. A., Finite-Amplitude Elastic-Plastic Wave Propagation in Fiber-Reinforced Composites, J. Appl. Phys., **45**, 4254–4261 (1974).
6 S. R. BODNER and PARTOM, Y., Constitutive Equations for Elastic-Viscoplastic Strain-Hardening Material, J. Appl. Mech., **42** 385–389 (1975).
7 A. MERZER and BODNER, S. R., Analytical Formulation of a Rate and Temperature Dependent Stress-Strain Relation, J. Eng. Mat. Tech., Trans. ASME **101**, 254–257 (1979).
8 Y. BENVENISTE and ABOUDI, J., A Nonlinear Mixture Theory for the Dynamic Response of a Laminated Composite under Large Deformations, J. Appl. Math. Phys. (ZAMP) **28**, 1067–1084 (1977).
9 S. R. BODNER, private commumication.

Eingereicht am 4. 9. 1980

Anschrift: Dr. JACOB ABOUDI, Tel-Aviv University, Dept. of Solid Mechanics, Materials and Structures, Ramat-Aviv, Tel-Aviv, 69978, Israel

ZAMM 61, 325 – 337 (1981)

H.-J. DIERSCH

Primitive Variable Finite Element Solutions of Free Convection Flows in Porous Media *)

A Galerkin-based finite element method is used to simulate two-dimensional steady and transient free convection flows in porous media driven by concentration gradients. For the present analysis pressure, velocities and concentration are chosen to be dependent variables. The resulting finite element equations and the iterative solution procedure are described. The analysis of several sample problems shows the validity and accuracy of the finite element model used.

Zur Simulation zweidimensionaler, stationärer und instationärer, freier Konvektionsströmungen in porösen Medien, hervorgerufen durch Konzentrationsgradienten, wird eine Galerkin-Finite-Element-Methode angewendet. Für das numerische Verfahren dienen Druck, Geschwindigkeiten und Konzentration als abhängige Variable. Eine Beschreibung der iterativen Näherungsmethode wird gegeben. Anhand der Berechnung verschiedener Testbeispiele zeigt sich die Gültigkeit und Genauigkeit des verwendeten Finite-Element-Modells.

Для симуляции двумерных стационарных и нестационарных свободных конвективных течений в пористых средах, вызванных градиентом концентрации, применяется метод конечных элементов Галеркина. Вычислительный метод использует давление, скорости и концентрацию в качестве зависимых переменных. Дано описание метода итеративных приближений. Пригодность и точность применямой модели конечных элементов показаны на тримере расчета различных примеров.

Introduction

Phenomena of free (natural) convection in porous media may occur in numerous geophysical and technical problems such as spreading of a pollutant in aquifers, utilizing of deep-lying groundwater, saltwater intrusion problems, heat transfer in geothermal systems or in fields of chemical engineering and some oil recovery techniques. In these situations fluid motion is driven or influenced by destabilizing density gradients producing buoyancy effects and resulting either from solute concentration or thermal gradients.

Considerable amount of theoretical and experimental work has been devoted to the study of such coupled transport processes in porous media which can often be idealized as saturated ones. A review of analytical, some numerical and experimental studies mainly concerning free thermal convection can be found in COMBARNOUS and BORIES [4] and CHENG [1]. Theoretical investigations of concentration induced convection in porous media as a modification or combination to thermal convection were presented by NIELD [19], RUBIN [21], [22], WEBER [30], RUBIN and CHRISTENSEN [22], KVERNVOLD and TYVAND [17].

To solve the governing conservation equations consisting of a nonlinear system of, at least, four (in two dimensions) simultaneous partial differential equations for mass, momentum, species concentration or energy without introducing particular simplifying assumptions or restrictions, one has to resort to numerical methods, particularly, when solving realistic problems. Amongst those methods finite difference approximations have been most popular (e.g., ELDER [7] [8], HORNE and O'SULLIVAN [15], HOLST and AZIZ [12]). Suitable difference schemes for the stream function POISSON equation (SOR [7], direct odd-even reduction method [15]) and for the transport equation (ARAKAWA explicit scheme [15]) were previously developed being particularly convenient for uniform rectangular meshes.

As an alternative to finite difference methods for solving the coupled transport-flow problem finite element methods were applied. In this field the finite element method was found to be well suited for solving more complex boundary value problems. As a similarity to free convection of viscous fluid investigated by TABARROK and LIN [27] who use a stream function-vorticity-temperature formulation or by GARTLING [9] who solves the equations in terms of primitive variables (i.e., pressure, velocity components, temperature), for free convection flows in porous media there are also, at least, two different solution strategies. CHENG [2] presented a stream function-concentration formulation via RAYLEIGH-RITZ's procedure for a steady state two-dimensional sea water encroachment problem. More recently, the same problem was alternatively solved by SEGOL et al. [25] for transient cases via a GALERKIN-based finite element technique using velocities, pressure and concentration as dependent variables. HUYAKORN and TAYLOR [16] introduced for these types of flows a further formulation using the hydraulic head and concentration as dependent variables and compared it numerically with a stream function-concentration and a primitive variable one alternatively used to the formulation presented by SEGOL et al. Some three-dimensional applications were already presented by TAYLOR and HUYAKORN [28] using a hydraulic head-concentration formulation. As result of the previous investigations the primitive variable formulation gives the best accurate values of velocities by providing a continuous representation of velocity field, which is desirable whenever the convective terms play a significant role in transport equation.

The purpose of the present paper is to describe and use a finite element model based on a primitive variable formulation for both steady and transient two-dimensional convection flows in porous media driven by concentration gradients in a more general manner as studied previously. Comparable previous works [16] [25] using initially

*) A revised version of a paper presented at the Int. Seminar "Čislennye metody rešenij uravnenij balansa" at Zentralinstitut für Mathematik und Mechanik of Akademie der Wissenschaften der DDR on Oct. 20—25, 1980 Berlin (GDR).

primitive variable formulations have restricted their approaches and applications to special sea water encroachment problems. It is of particular interest here to study the behaviour of the finite element method under definitive varying physical conditions characterizing the convection process.

Basic equations

The equations governing the transient natural convective motion in a porous medium are (summation convention)

$$L_1(v_i) = v_{i,i} = 0 \tag{1}$$

the conservation of fluid mass (continuity equation),

$$L_2(p, v_i, C) = v_i + \frac{k_{ij}}{\mu\alpha}(p_{,j} + \varrho g_j) = 0 \tag{2}$$

the conservation of fluid momentum (generalized DARCY equation),

$$L_3(v_i, C) = \frac{\partial C}{\partial t} + v_i C_{,i} - (D_{ij}C_{,j})_{,i} = 0 \tag{3}$$

the conservation of species concentration (transport equation) and

$$\varrho = \varrho_0(1 + \varepsilon C/C_1) \tag{4a}$$

with

$$\varepsilon = [\varrho(C_1) - \varrho_0]/\varrho_0 \tag{4b}$$

equation of state of fluid's density, where v_i are the components of the averaged pore velocity (LT^{-1}), k_{ij} are components of permeability tensor (L^2), μ is the dynamic viscosity of fluid ($ML^{-1}T^{-1}$), α is the effective porosity of porous medium, p is the fluid pressure ($ML^{-1}T^{-2}$), ϱ is the fluid density (ML^{-3}), g_j are the components of gravity vector (LT^{-2}) in the direction x_j, C is the species concentration (ML^{-3}), D_{ij} are the components of the dispersion tensor (L^2T^{-1}), ε is the density difference ratio, ϱ_0 is the reference fluid density (ML^{-3}) and C_1 is the maximum value of concentration (ML^{-3}).

Inherent in the derivation of these equations is a number of assumptions (DIERSCH [5]), mainly

- The porous medium is saturated and isothermal.
- The BOUSSINESQ assumption.
- Inertial effects are small (low REYNOLDS number).
- Viscosity and porosity are constant.
- The fluid is a binary miscible system (chemical compound represents a solute).
- No chemical reactions and adsorption take place.

For the above equations the initial and boundary conditions required are

$$v_i = \hat{v}_i \quad \text{on } B_1\,, \qquad p = \hat{p} \quad \text{on } B_2\,, \tag{5a), (5b}$$

$$C(x_i, 0) = C_0\,, \qquad C = \hat{C} \quad \text{on } B_3\,, \tag{5c), (5d}$$

$$D_{ij}C_{,j}l_i + q_n = 0 \quad \text{on } B_4 \tag{5e}$$

where l_i are the directional cosines of the outward unit normal vector, $B_1 \cup B_2 = B$, $B_3 \cup B_4 = B$, B is the entire boundary and q_n is the prescribed normal concentration flux.

Finite Element Formulation (p—v_1—v_2—C—model)

The finite element process has been thoroughly described by several authors (e.g., see [31]) and only details are presented here in context of the interesting class of problems.

To obtain a continuous velocity field the finite element model used solves simultaneously pressure and velocities. Such a procedure is similar to that proposed by MEISSNER [18]. Therefore, the approximation of the non-linear system of Eqs. (1)—(3) leads to two sets of equations, where the coupled transport equation is evaluated separately.

The flow domain, A, of interest with the boundary, B, is divided into J subdomains, A^e, called finite elements, each bounded by B^e. These elements are assumed to be fixed in space due to the EULERian description of the fluid motion (Eqs. (1)—(3)). Within each element, e, the dependent variables p, v_i, and C are interpolated as continuous functions p^e, v_i^e, C^e of compatible order in form of the polynomials

$$p^e(x_i) = M_l^e(x_i)\,P_l^e\,, \qquad v_i^e(x_i) = N_k^e(x_i)\,V_{ik}^e\,, \qquad C^e(x_i, t) = N_k^e(x_i)\,C_k^e(t)\,, \tag{6a), (6b), (6c}$$

$l = 1, 2, \ldots, r$; $k = 1, 2, \ldots, s$, so that the pressure, velocity and concentration fields for the entire domain; A can be approximated by

$$p(x_i) \simeq \bigcup_{e=1}^{J} p^e(x_i) \quad \text{etc.} \tag{7}$$

the union of the element functions (6) satisfying the essential boundary conditions (5) and having the properties required

$$M_n^e(x_{im}),\ N_n^e(x_{im}) = \delta_{nm} \quad \text{for} \quad x_{im} \in A^e\,, \tag{8a}$$

$$M_n^e(x_i),\ N_n^e(x_i) = 0 \quad \text{for} \quad x_i \notin A^e\,, \tag{8b}$$

where $M_n^e(x_i)$, $N_n^e(x_i)$ are different coordinate (basis) functions called element shape functions associated with selected points n called nodes, P_n^e, V_{in}^e and $C_n^e(t)$ are coefficients at nodes n yet to be determined (nodal unknowns), x_{im} are the space coordinates of the mth node, δ_{nm} is the KRONECKER delta.

Substituting Eqs. (6) and (7) into Eqs. (1)—(3) it must be required that the resulting residuals (errors) tend to zero in a weighted sense (method of weighted residuals), i.e.

$$\int_A W_n L_1(N_k, V_{ik})\, dA = 0\,, \qquad n = 1, 2, \dots, R\,, \tag{9a}$$

$$\int_A X_m L_2(M_l, N_k, P_l, V_{ik}, C_k)\, dA = 0\,, \qquad m = 1, 2, \dots, S\,, \tag{9b}$$

$$\int_A X_m L_3(N_k, V_{ik}, C_k)\, dA = 0\,, \tag{10}$$

where $N_k = \bigcup_{e=1}^{J} N_k^e$ etc., W_n and X_m are suitable weighting functions to be prescribed.

After applying GREEN's theorem to second order derivative terms the resulting approximate equations are written in matrix form

$$[\boldsymbol{H}]\{\boldsymbol{\pi}\} + \{\boldsymbol{G}\} = 0\,, \tag{11a}$$

$$[\boldsymbol{S}]\{\boldsymbol{C}(t)\} + [\boldsymbol{O}]\{d\boldsymbol{C}(t)/dt\} + \{\boldsymbol{F}(t)\} = 0 \tag{12a}$$

where the arrays are given by

$$\boldsymbol{H}_{kl} = \sum_e \iint_{A^e} \begin{bmatrix} 0 & W_k^e N_{l,1}^e & W_k^e N_{l,2}^e \\ \frac{k_{11}}{\mu\alpha} X_k^e M_{l,1}^e & X_k^e N_l^e & 0 \\ \frac{k_{22}}{\mu\alpha} X_k^e M_{l,2}^e & 0 & X_k^e N_l^e \end{bmatrix} dA \tag{11b}$$

where it is further assumed that $k_{ij} = 0$ for $i \neq j$,

$$\boldsymbol{\pi}_k = \begin{Bmatrix} P_k \\ V_{1k} \\ V_{2k} \end{Bmatrix} \text{ having } \quad P_k = \bigcup_e P_k^e \quad \text{etc.}\,, \tag{11c}$$

$$\boldsymbol{G}_k = \sum_e \iint_{A^e} \begin{Bmatrix} 0 \\ \frac{k_{11}}{\mu\varepsilon} \varrho_0 g_1 X_k^e \left[1 + \frac{\varepsilon}{C_1} \sum_{\nu=1}^{s} (N_\nu^e C_\nu^e(t))\right] \\ \frac{k_{22}}{\mu\varepsilon} \varrho_0 g_2 X_k^e \left[1 + \frac{\varepsilon}{C_1} \sum_{\nu=1}^{s} (N_\nu^e C_\nu^e(t))\right] \end{Bmatrix} dA\,, \tag{11d}$$

$$\boldsymbol{S}_{kl} = \sum_e \iint_{A^e} \left[D_{ij} X_{k,i}^e N_{l,j}^e + X_k^e \sum_{\nu=1}^{s} (N_\nu^e V_{i\nu}^e)\, N_{l,i}^e\right] dA\,, \tag{12b}$$

$$\boldsymbol{O}_{kl} = \sum_e \iint_{A^e} X_k^e N_l^e\, dA\,, \qquad \boldsymbol{F}_k(t) = -\sum_e \int_{B^e} X_k^e D_{ij} \sum_{\nu=1}^{s} (N_{\nu,j}^e C_\nu^e(t))\, l_i\, ds\,. \tag{12c), (12d}$$

Such an assembly process enforces to meet inter-element continuity of approximate pressure, velocity and concentration. Involving at most first order derivatives occurred in the above finite element equations, C_0-type finite elements can be applied performing the conformity condition required for the approximate functions.

According to the character of the flow equations approximated by Eqs. (11) it has been shown [13] [16] that the order of the polynomial expansion for the pressure should be one lower than the expansions for the velocity components. For the present study eight-node isoparametric finite elements of SERENTIPITY type [31] are used, where the velocity components and concentration are approximated biquadratically ($s = 8$) within each element, while the pressure varies bilinearly ($r = 4$). Similarly, the element geometry is defined biquadratically by

$$x_i^e = N_m^e x_{im}^e\,, \qquad m = 1, 2, \dots, 8\,. \tag{13}$$

Therefore, the global derivatives of an element shape function, e.g., $N_l^e(\xi, \eta)$, are given by

$$\begin{Bmatrix} N_{l,1}^e \\ N_{l,2}^e \end{Bmatrix} = [\boldsymbol{J}]^{-1} \begin{Bmatrix} \frac{\partial N_l^e}{\partial \xi} \\ \frac{\partial N_l^e}{\partial \eta} \end{Bmatrix}, \tag{14a}$$

where $[\boldsymbol{J}]$ is the JACOBIan matrix written as

$$[\boldsymbol{J}] = \begin{bmatrix} \frac{\partial x_1^e}{\partial \xi} & \frac{\partial x_2^e}{\partial \xi} \\ \frac{\partial x_1^e}{\partial \eta} & \frac{\partial x_2^e}{\partial \eta} \end{bmatrix} = \begin{bmatrix} \frac{\partial N_m^e}{\partial \xi} x_{1m}^e & \frac{\partial N_m^e}{\partial \xi} x_{2m}^e \\ \frac{\partial N_m^e}{\partial \eta} x_{1m}^e & \frac{\partial N_m^e}{\partial \eta} x_{2m}^e \end{bmatrix} \qquad m = 1, 2, \dots, 8 \tag{14b}$$

to map the physical space of each element onto a standard element in a local coordinate system (ξ, η). Then, integrations required to generate the matrices (vectors) $\boldsymbol{H}$, $\boldsymbol{G}$, $\boldsymbol{S}$, $\boldsymbol{O}$ and $\boldsymbol{F}$ are performed in the (ξ, η)-system using GAUSSian quadrature within the limits ± 1, where

$$\mathrm{d}A = \det [\boldsymbol{J}]\, \mathrm{d}\xi\, \mathrm{d}\eta \tag{14c}$$

and

$$\mathrm{d}s = (\mathrm{d}x_i^e\, \mathrm{d}x_i^e)^{\frac{1}{2}} \tag{14d}$$

with

$$\mathrm{d}x_i^e = \frac{\partial N_m^e}{\partial \xi} x_{im}^e\, \mathrm{d}\xi + \frac{\partial N_m^e}{\partial \eta} x_{im}^e\, \mathrm{d}\eta\,, \qquad m = 1, 2, \ldots, 8\,.$$

Galerkin Method, Time Discretization, and Some Properties

Within the finite element context the GALERKIN method gives usually the best approximations having, in particular interest here, small or negligible truncation errors associated with the spatial derivatives when higher order basis functions are used [29]. However, in non-self-adjoint problems included here the approximate solution can contain spurious oscillations in situations where first order derivative (convective) terms are dominant. Alternatively, as suggested by CHRISTIE et al. [3], HEINRICH et al. [11], the equations may be artificially damped by using special non-symmetric weighting functions (upwinding), albeit with a loss of accuracy. To avoid additional computations and errors inherent in the upwind process, especially when higher order basis functions and transient problems are to be solved, in the present study standard GALERKIN weighting within the approximate equations (11) and (12) is employed. Therefore, assuming biquadratic and bilinear approximations of the velocity, concentration and pressure, respectively, the weighting functions are

$$W_k^e = M_k^e; \qquad X_k^e = N_k^e\,. \tag{15}$$

That means, considering a quasi-one-dimensional problem in x_1-direction, that the GALERKIN-based finite element solution of the transport equation in terms of discrete C_m $(m = 1, 2, \ldots)$ is oscillation free for (see [3]) all

$$\frac{v_1 a}{D} > 0 \tag{16}$$

with a fourth order accuracy, if m corresponds to a corner node in an element, and

$$\frac{v_1 a}{D} < 4 \tag{17}$$

with a second order accuracy, if m corresponds to a mid-side node in an element, where D, v_1 and a represent the constant dispersion coefficient, the velocity component and the element length, respectively. However, the restricted condition (17) for the mid-side nodes does not exist when a natural type boundary condition at the downstream boundaries can be specified and the solutions are non-oscillatory in all cases [10] [32].

To solve the ordinary differential equation (12a) in time a simple finite difference approximation is used which leads to the two-level recurrence scheme over the time interval $(t, t + \Delta t)$ as

$$(\boldsymbol{O}/\Delta t + \boldsymbol{S}\sigma)\, \boldsymbol{C}(t + \Delta t) = [\boldsymbol{O}/\Delta t - \boldsymbol{S}(1 - \sigma)]\, \boldsymbol{C}(t) - [\sigma \boldsymbol{F}(t + \Delta t) + (1 - \sigma)\, \boldsymbol{F}(t)]\,, \tag{18}$$

where Δt and σ are the time step and a time weighting factor, respectively. Specifying weighting factors one obtains several time marching schemes, in particular

for $\sigma = 1/2$ central difference (CRANK-NICOLSON scheme)
for $\sigma = 1$ backward difference (fully implicit scheme).

While the CRANK-NICOLSON scheme is of second order in time $O(\Delta t^2)$, the other schemes provide only first order accuracy $O(\Delta t)$ .Therefore, the lower order schemes involve influences of time truncation errors (e.g., [6] concerning linear problems), known as numerical dispersion, in a greater extent and increase the real dispersion. For instance, considering the transport in the x_1-direction the error estimate holds

$$D = D_{\text{real}} + \Delta t v_1^2/2\,. \tag{19}$$

Furthermore, it is well known that for $\sigma \geqq 1/2$ the schemes satisfying the expression derived from (18)

$$\omega_i \Delta t(2\sigma - 1) > -2 \tag{20}$$

are unconditionally stable, i.e., for all Δt and ω_i, in which $\omega_i = S_i/O_i$ correspond to the modal eigenvalues of the system (18). In order to avoid oscillations, however, the schemes have to require

$$(1 - \sigma)\, \omega_i \Delta t < 1\,, \tag{21}$$

i.e., for $\sigma < 1$ there are restricted eigenvalues or time steps. From above only the fully implicit scheme $(\sigma = 1)$ gives non-oscillatory results for all ω_i and Δt.

Solution Procedure

The finite element equations (11), (12) and (18) are coupled by buoyancy forces and their terms $\boldsymbol{H}$ and $\boldsymbol{S}$ are unsymmetric and banded, while the term $\boldsymbol{O}$ is symmetric and banded. Writing the resulting equations (11) and (18) to be solved symbolically as

$$\boldsymbol{H} \cdot \boldsymbol{\pi}_{t+\Delta t} + \boldsymbol{G}(\boldsymbol{C}_{t+\Delta t}) = 0\,, \qquad \boldsymbol{D}(\boldsymbol{V}_{t+\Delta t}) \cdot \boldsymbol{C}_{t+\Delta t} = \boldsymbol{E} \cdot \boldsymbol{C}_t + \hat{\boldsymbol{F}}_t\,, \tag{22), (23}$$

where the matrix $\boldsymbol{E}$ is generally also velocity dependent, i.e., $\boldsymbol{E} = \boldsymbol{E}(\boldsymbol{V}_t)$, however, for a fully implicit time marching scheme $\boldsymbol{E}$ becomes linear, i.e., $\boldsymbol{E} = \boldsymbol{O}/\Delta t$, the solution of this non-linear system may be performed in the following iterative procedure as evidenced by Fig. 1. For the time level $t + \Delta t$ the iteration written as

$$\begin{aligned} &\boldsymbol{H} \cdot \boldsymbol{\pi}^{\tau}_{t+\Delta t} + \boldsymbol{G}(\boldsymbol{C}^{\tau}_{t+\Delta t}) = 0 \\ &\boldsymbol{D}(\boldsymbol{V}^{\tau}_{t+\Delta t}) \cdot \boldsymbol{C}^{\tau+1}_{t+\Delta t} = \boldsymbol{E} \cdot \boldsymbol{C}_t + \hat{\boldsymbol{F}}_t \\ &\boldsymbol{H} \cdot \boldsymbol{\pi}^{\tau+1}_{t+\Delta t} + \boldsymbol{G}(\boldsymbol{C}^{\tau+1}_{t+\Delta t}) = 0 \\ &\vdots \\ &\text{etc.} \end{aligned} \tag{24}$$

alternates between the two equations, where the superscript indicates the iteration level. In Eq. (24) the first step of solving the flow equations is only necessary when free and forced (mixed) convection is taken into consideration, otherwise the velocity field at the beginning of the iteration stems from the previous time level, i.e., $\boldsymbol{V}^0_{t+\Delta t} = \boldsymbol{V}^{\tau}_t$.

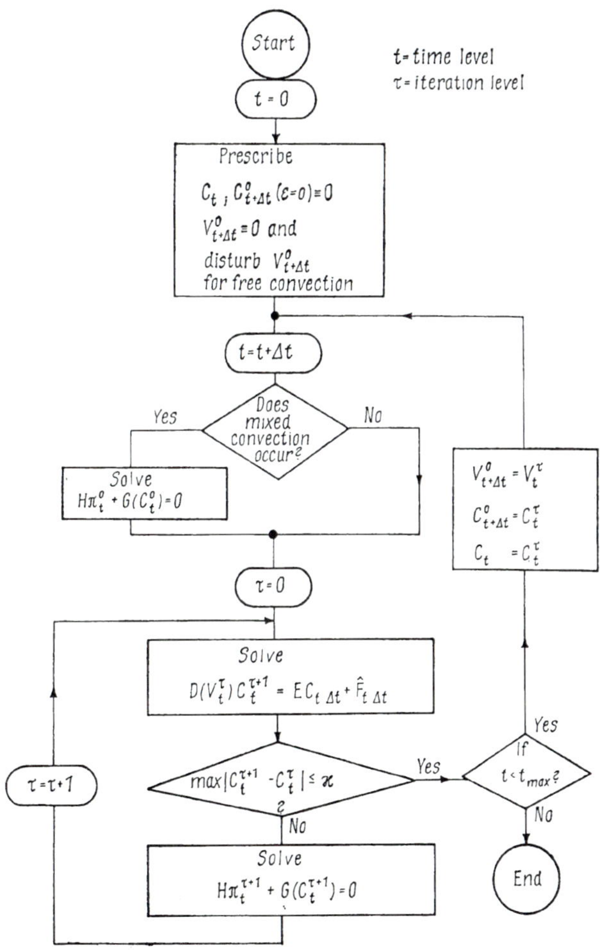

Fig. 1. Flow diagram of the iterative solution procedure used

For computing steady state convection flows only the inner iteration loop (Fig. 1) should be carried out and the transient term in Eq. (18) or (23) has to be suppressed.

The unsymmetric matrix equations (22) and (23) are solved by specialized Gaussian elimination techniques. On the one hand, the solution of the set of algebraic equations (22) is performed by a frontal algorithm with limited pivoting [14]. On the other hand, the equations of (23) are solved by a profile Crout elimination scheme without pivoting fully exploiting the system structure [31]. The evaluation of the integrals in (11) and (12) is through Gaussian quadrature using a 3 × 3 point integration scheme.

The calculations were executed on a BESM-6 computer with a 40-bit mantissa for single precision, where 28,000 decimal words of core storage are available. This core version accommodates up to 100 elements of the above type.

Sample Problems

In order to verify the above finite element model typical problems of steady and transient free convection already previously analyzed by other methods are studied. The examples of those problems refer to cellular convective motions in porous layers of rectangular geometries generally concerning free thermal convection, where effects of mechanical dispersion are negligible. Here the convection process in a porous layer can be unified and characterized by the (filtration) RAYLEIGH number defined as

$$Ra = \frac{(\varepsilon/C_1)\, g\, \Delta Ckh}{\nu D} \tag{25}$$

involving the simplifications in Eqs. (2) and (3) as $k_{ij} = k\delta_{ij}$ and $D_{ij} = D\delta_{ij}$, respectively, where k is a constant permeability coefficient, D is a constant dispersion (diffusion) coefficient, g is the gravitational acceleration, ΔC is the concentration difference between the two surfaces of the porous layer having the depth h and ν ($= \mu/\varrho_0$) is the kinematic viscosity. According to stability analysis [1] [4] [19] it can be shown for this type of problems that the fluid remains stationary below a certain critical value of the RAYLEIGH number (Ra_{crit1}) with mass (or heat) transfer by diffusion only. Above Ra_{crit1} ($4\pi^2$ as standard criterion for the onset of free convection; furthermore, however, depending on the layer (cell) geometry and boundary conditions) the flow moves in steady convection cells or two-dimensional rolls until the RAYLEIGH number reaches a second critical value (Ra_{crit2}) at which transient disturbances first appear, called the fluctuating convective state. This value lies in the range 240–390 [1].

Steady cases

Unicellular convection. For this case the convection process is well understood and many results exist. In the present study the free convective currents in a quadratic region (i.e., $h = b$, where b is the cell width) are computed for various RAYLEIGH numbers. It is well known that here an unicellular motion occurs for $Ra \geqq 4\pi^2$. Two different finite element meshs are selected as shown in Fig. 2 to investigate the influences of accuracy and

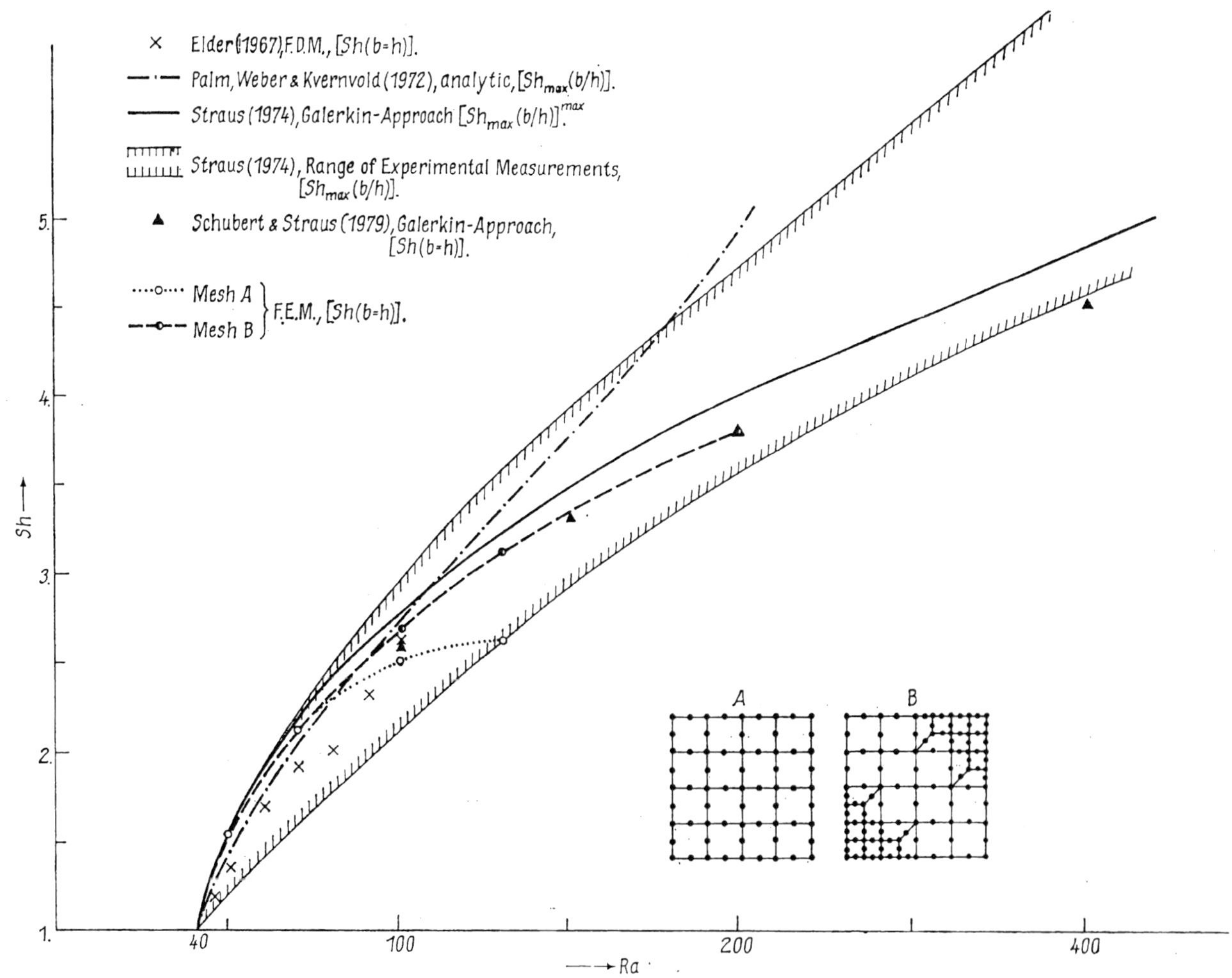

Fig. 2. Finite element meshes used and SHERWOOD numbers Sh computed for various RAYLEIGH numbers Ra in comparison with other results (Note that several solutions are based on maximum SHERWOOD numbers being a function of cell aspect ratio $b/h \leqq 1$. However, in the present range of Ra it can be assumed that Sh_{max} occurs for $b/h \simeq 1$)

stability. Following boundary conditions have to be satisfied:

$$v_2 = 0\,, \quad C = 1 \quad \text{on the upper boundary}\,,$$

$$v_2 = 0\,, \quad C = 0 \quad \text{on the lower boundary}\,,$$

$$v_1 = 0\,, \quad \frac{\partial C}{\partial n} = 0 \quad \text{on all vertical boundaries}\,.$$

Due to confinement by the boundaries the pressure remains to be indeterminate and it is necessary to specify the pressure at one point in the flow domain (e.g., at the centre as used in the present examples). The computations were performed at RAYLEIGH numbers of 50, 70, 100, 130, 200. The SHERWOOD number similar to the NUSSELT number applied to thermal convection is defined as

$$Sh = \frac{h}{bC_1} \int\limits_0^b \frac{\partial C(x_1, 0)}{\partial x_2}\, dx_1 \tag{26}$$

to indicate the mean mass transfer through the cell. In Fig. 2 the computed SHERWOOD numbers Sh are compared with other solutions [7] [20] [24] [26]. It is found the relatively coarse mesh A gives results being in good agreement with previous studies, however, only below a RAYLEIGH number of 100. Above 70 the solutions fall and become unstable at $\mathrm{Ra} = 200$. In using the mesh B only slightly refined in the two corners to adopt the high velocity gradients there the computed SHERWOOD numbers are in excellent agreement up to 200. For these cases the concentration distributions obtained are shown in Fig. 3.

The convergence of the solutions is indicated in Fig. 4 showing that the rate of convergence is more rapid at higher values of Ra and that for $Ra > 70$ there is some overshoot. This behaviour is similar to the finite difference solutions obtained by ELDER [7]. In the present analysis the applied convergence tolerance defined in Fig. 1 as

$$\max |C_i^{\tau+1} - C_i^{\tau}| \leqq \varkappa \tag{27}$$

was $\varkappa = 10^{-4}$, while ELDER used 10^{-3}. Nevertheless the finite element technique proposed is up to four times faster than ELDER's computations.

Multicellular convection. The computed flow pattern of a more complex tricellular convection problem is presented in Fig. 5. Here the boundary conditions were

$$v_2 = 0\,, \quad C = 1 \quad \text{on } AB\,, \quad v_2 = 0\,, \quad \frac{\partial C}{\partial n} = 0 \quad \text{on } BC\,,$$

$$v_2 = 0\,, \quad C = 0 \quad \text{on } DE\,, \quad v_1 = 0\,, \quad \frac{\partial C}{\partial n} = 0 \quad \text{on } AE \text{ and } CD\,.$$

According to the assumed length d $(= AB/AC) = 0.769$ the resulting flow pattern as plotted in Fig. 5 has to be tricellular. For the present example the computations were terminated when values of concentration change by less than 0.1%. This convergence criterion was reached after 14 iterations. The corresponding concentration distribution is illustrated in Fig. 6.

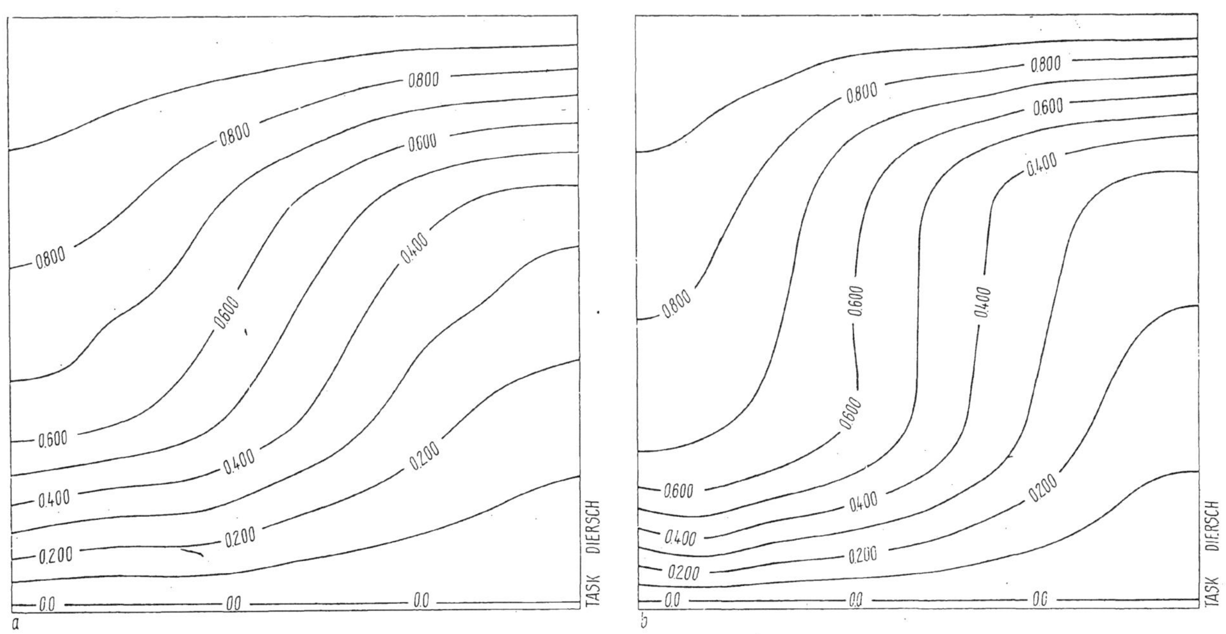

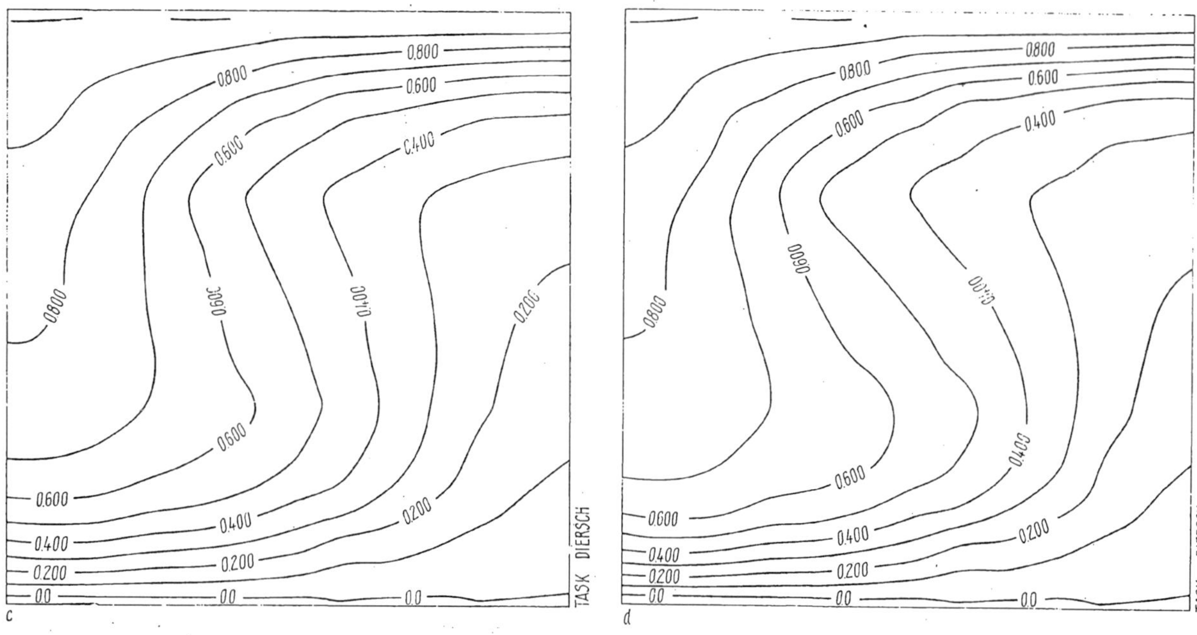

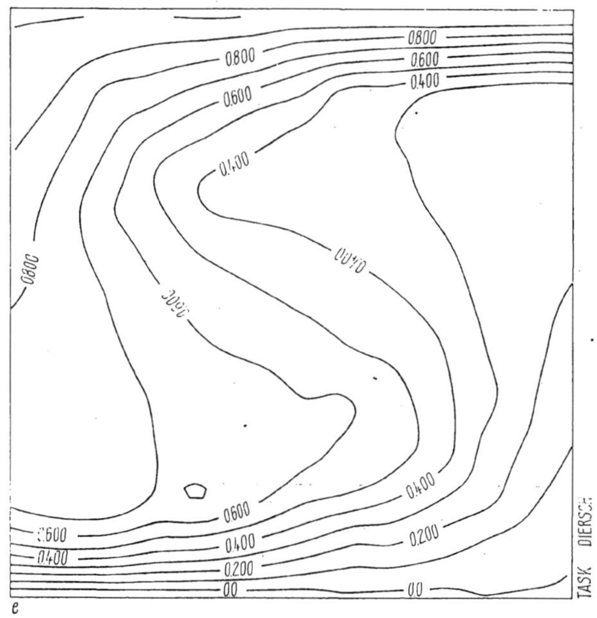

Fig. 3. Iso-concentration lines computed at Rayleigh numbers $Ra = 50$ (Fig. a), 70 (Fig. b), 100 (Fig. c), 130 (Fig. d), 200 (Fig. e)

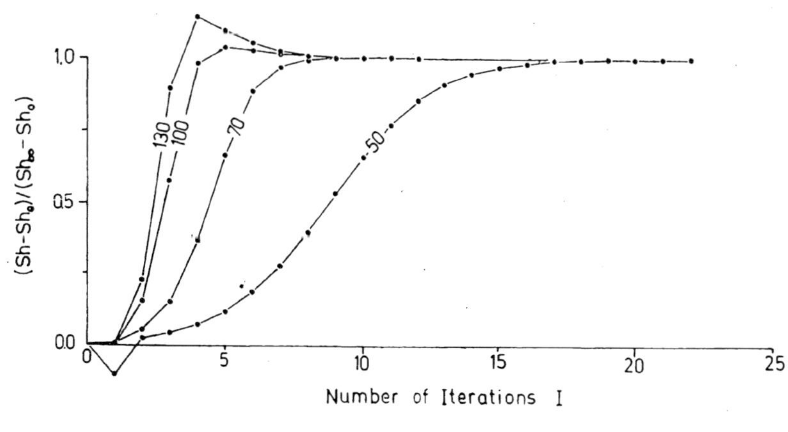

Fig. 4. Convergence behaviour for different Rayleigh numbers (the reference Sherwood number $Sh_0 = 1$ for mass transport without convection, the final Sherwood number Sh_∞ raeched at convergence tolerance $\varkappa = 10^{-4}$)

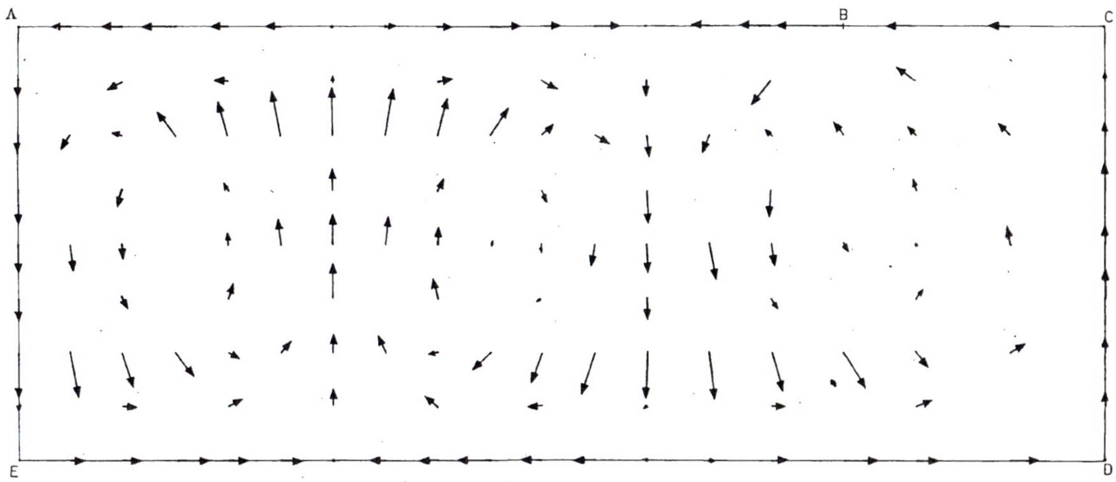

Fig. 5. Computed velocity field of tricellular convection for $Ra = 80$, d (= AB/AC) = 0.769, $\varkappa = 10^{-3}$ (total number of nodes = 135, total number of degrees-of-freedom = 455)

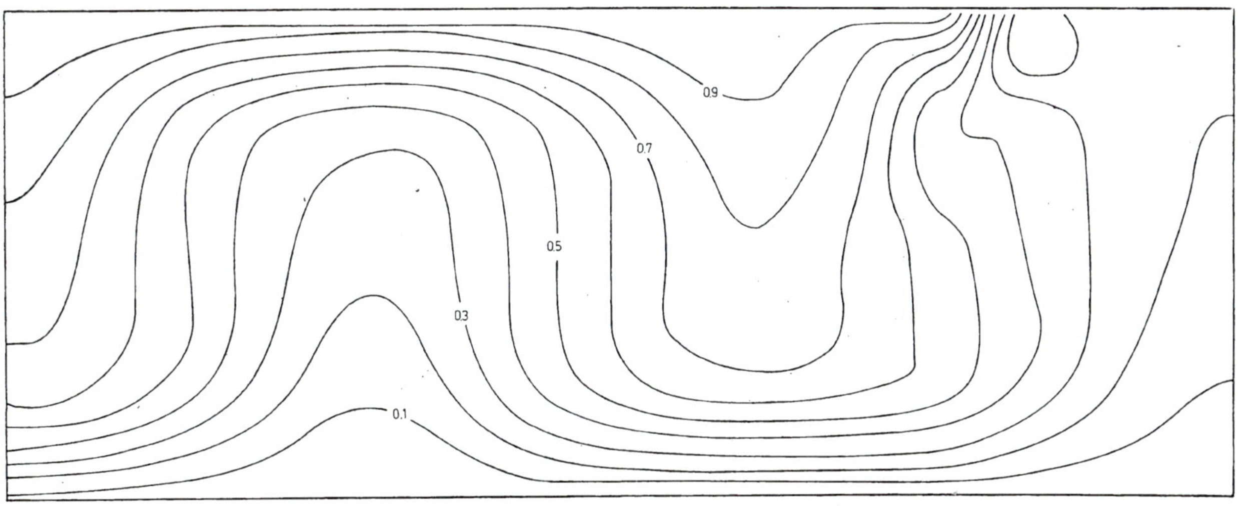

Fig. 6. Computed iso-concentration lines of tricellular convection for $Ra = 80$, $d = 0.769$, $\varkappa = 10^{-3}$

Transient case

To compare the finite element solutions with previous findings for transient convections a typical model problem already investigated by ELDER [8] and HORNE and O'SULLIVAN [15] who utilized finite difference methods is considered. Figure 7 shows the symmetric half of the flow domain idealized by 63 finite elements in the example finally solved. It should be pointed out that a coarser mesh was not sufficient to simulate especially the starting phase in the upper region. For this example the following boundary conditions hold

$$v_2 = 0\,, \quad C = 1 \quad \text{on } AB\,, \quad v_2 = 0\,, \quad C = 0 \quad \text{on } BC \text{ and } DE\,,$$

$$v_1 = 0\,, \quad \frac{\partial C}{\partial n} = 0 \quad \text{on } CD \text{ and } AE\,.$$

Moreover, the initial condition is

$$C(x_i, t) = 0 \quad \text{at} \quad t = 0\,.$$

The problem was solved by a fully implicit time marching scheme. Within the time steps the solutions for $C(x_i, t)$ were determined to a precision of 3%. Introducing the dimensionless time T defined as

$$T = \frac{D}{h^2}\,t \tag{28}$$

the computed velocity and concentration distributions for $Ra = 400$ at selected times are shown in Fig. 8.

At first a relatively thin layer is formed in the upper part. The first motion is an eddy at the end of the concentration input boundary (point B). This end eddy transports the concentration away from the end of the boundary and produces a small bump in the concentration lines. In this time a small eddy of reverse circulation is associated with the end eddy (see velocities for $T = 0.00625$ in Fig. 8). At the time $T = 0.01$ this double eddy is well established and produces a small rapidly descending column of fluid. However in addition, a small eddy forms near

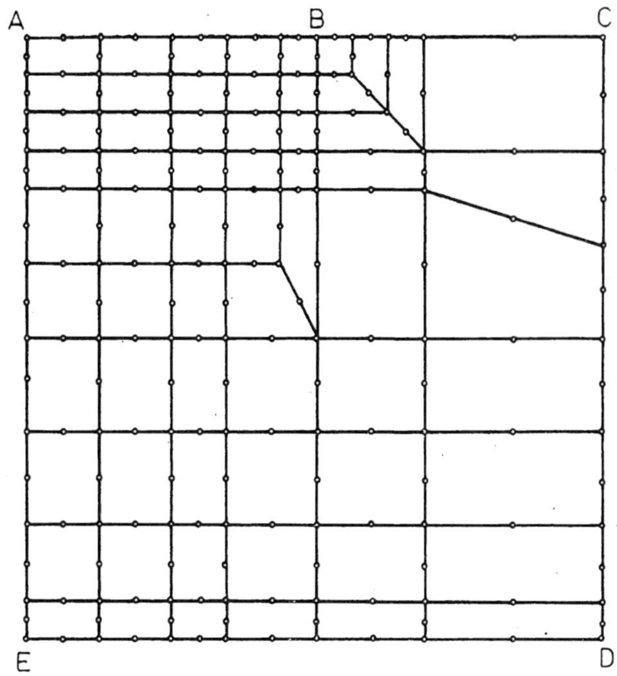

Fig. 7. Finite element mesh for simulating transient free convection ($d = 0.5$, total number of nodes = 222, total number of degrees-of-freedom = 746)

the top of the axis of symmetry. Now, this eddy grows rapidly. At the time $T = 0.01438$ three eddies are established, where the upper two eddies having the same sense of rotation begin to fuse. This fusion continues to increase (see $T = 0.01688$) and at the time $T = 0.02$ the eddies have now merged into a single large eddy. Now, the single eddy grows more slowly to a nearly steady state.

For the analysis time steps of different length were used in which 3 to 10 iterations were necessary to satisfy the convergence criterion above mentioned. The agreement with ELDER's results is satisfactory. A further discussion of the results will be included in a subsequent paper.

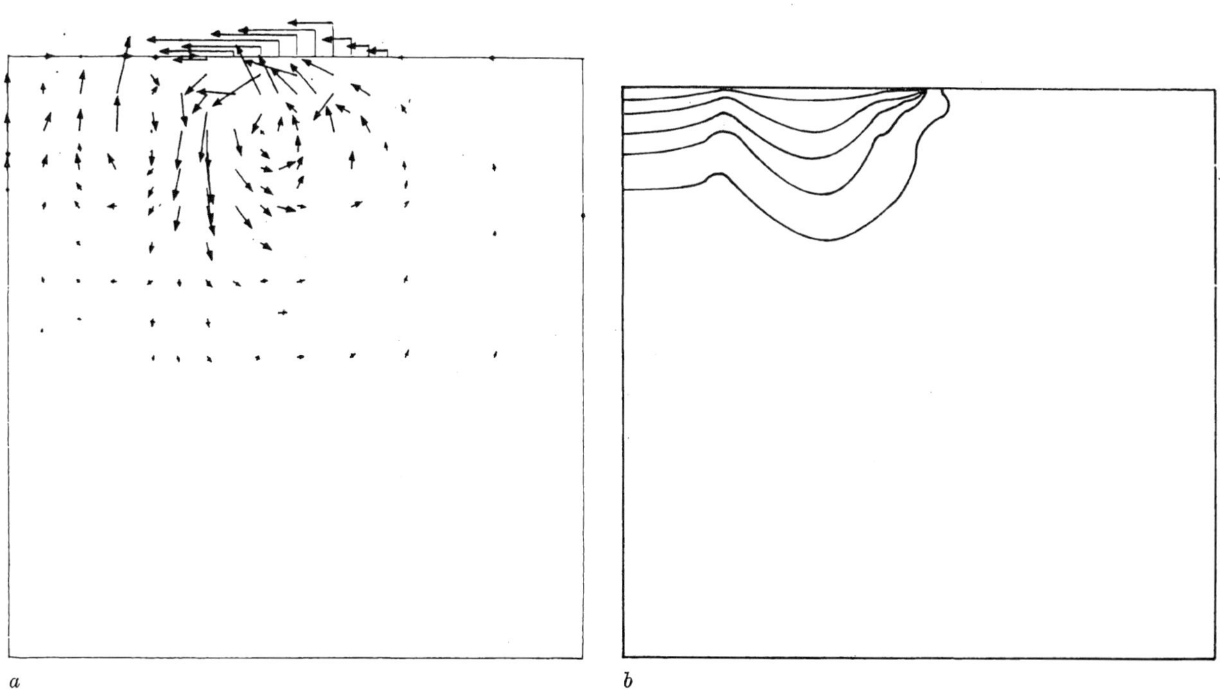

a *b*

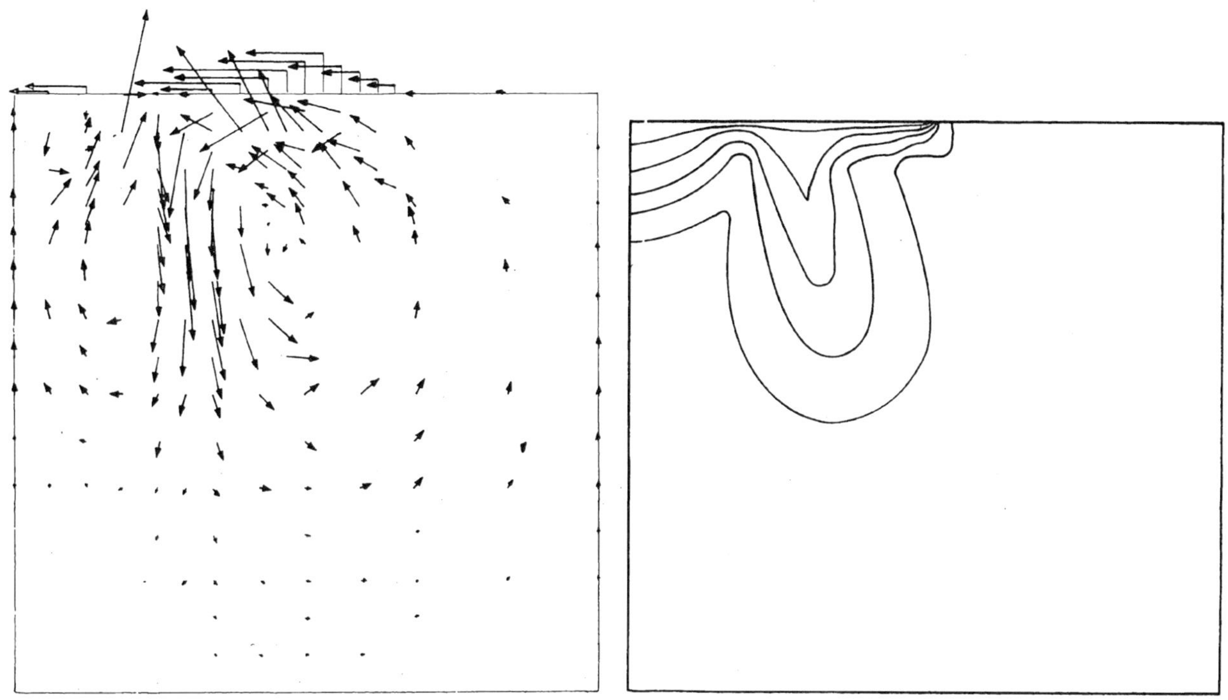

c d

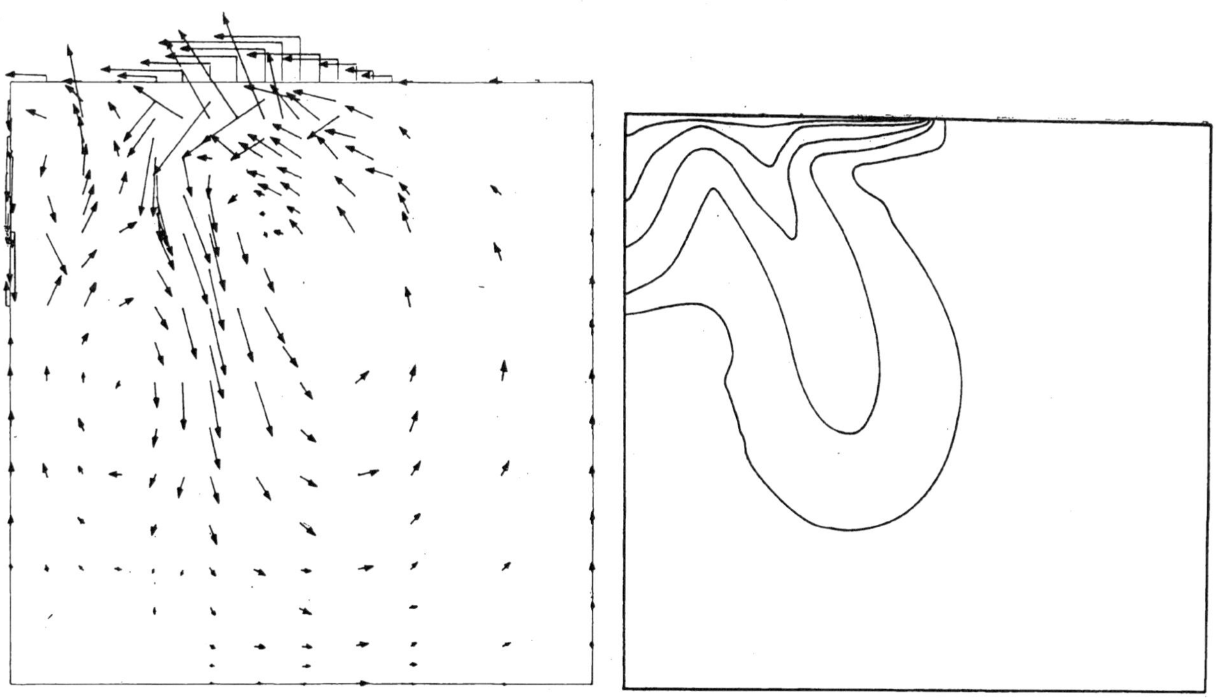

e f

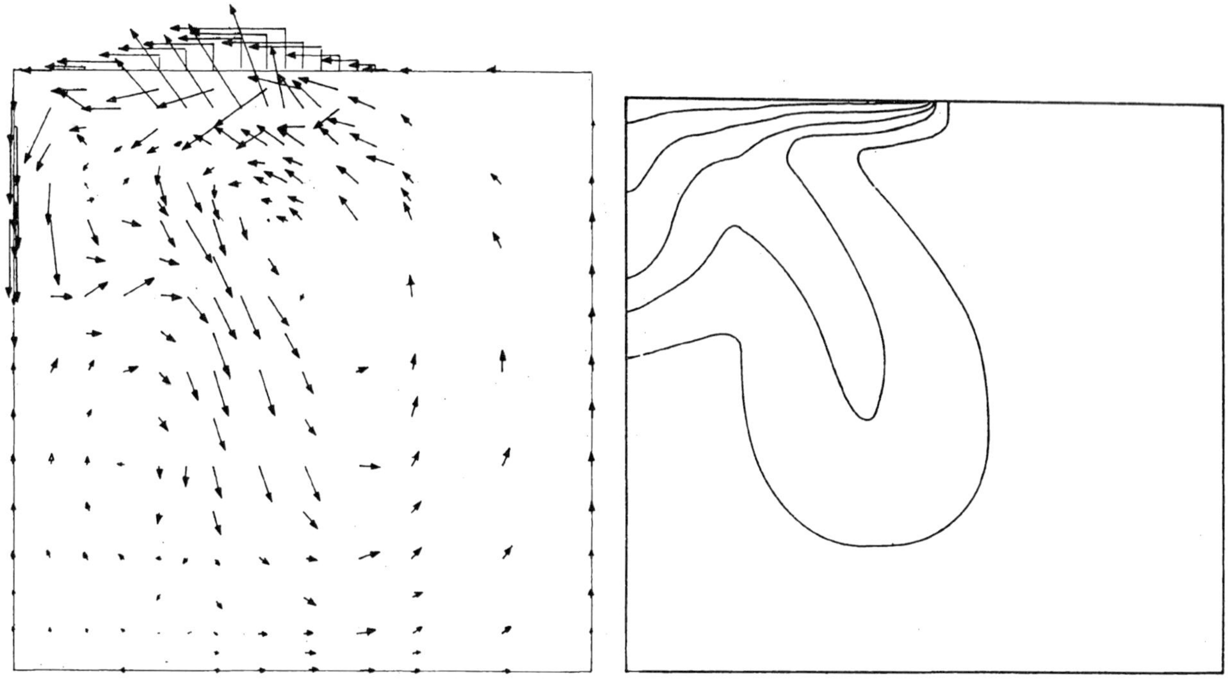

g h

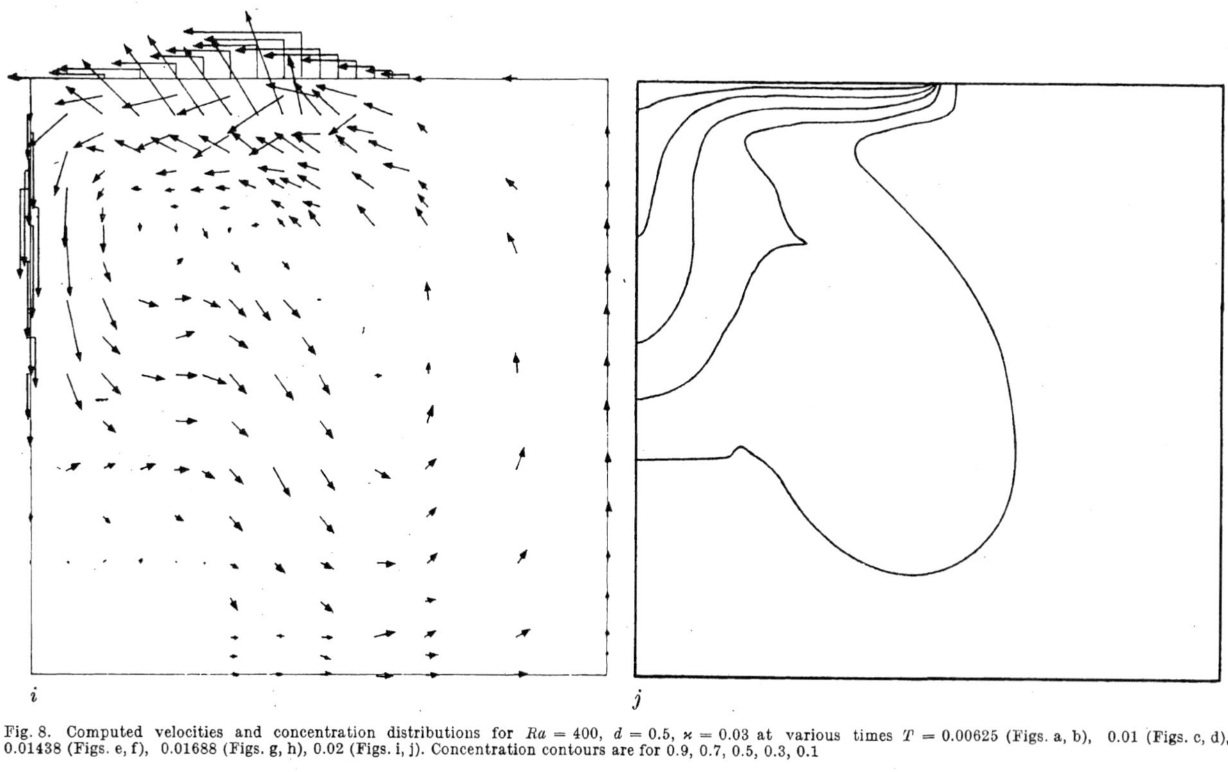

i j

Fig. 8. Computed velocities and concentration distributions for $Ra = 400$, $d = 0.5$, $\varkappa = 0.03$ at various times $T = 0.00625$ (Figs. a, b), 0.01 (Figs. c, d), 0.01438 (Figs. e, f), 0.01688 (Figs. g, h), 0.02 (Figs. i, j). Concentration contours are for 0.9, 0.7, 0.5, 0.3, 0.1

References

1 CHENG, P., Heat transfer in geothermal systems, in: Advances in heat transfer, Academic Press, New York/San Francisco/London, **14** (1978), S. 1–105.

2 CHENG, R. T., On the study of convective dispersion equation, in: ODEN, J. T. / ZIENKIEWICZ, O. C. / GALLAGHER, R. H. / TAYLOR, C. (Ed.), Finite Element Methods in flow problems, Proc. Int. Symp., Swansea 7.–11. 1. 74, Huntsville, Alabama: UAH Press 1974, S. 29–47.

3 CHRISTIE, I.; GRIFFITHS, D. F.; MITCHELL, A. R.; ZIENKIEWICZ, O. C., Finite element methods for second order differential equations with significant first derivatives, Int. J. Num. Meth. Engng. **10** (1976), S. 1389–1396.

4 COMBARNOUS, M. A.; BORIES, S. A., Hydrothermal convection in saturated porous media, in: Advances in hydroscience, Academic Press, New York/San Francisco/London, **10** (1975), S. 231–307.
5 DIERSCH, H.-J., Mehrphasiger Stoff-, Impuls- und Wärmetransport im ungesättigten und gesättigten porösen Medium, Acta Hydrophysica **25** (1980) 4, S. 251–407.
6 DIERSCH, H.-J., Finite-Element-GALERKIN-Modell zur Simulation zweidimensionaler konvektiver und dispersiver Stofftransportprozesse im Boden, Acta Hydrophysica **26** (1981) 1, S. 5–44.
7 ELDER, J. W., Steady free convection in a porous medium heated from below, J. Fluid Mech. **27** part 1 (1967), S. 29–48.
8 ELDER, J. W., Transient convection in a porous medium, J. Fluid Mech. **27**, part 3 (1967), S. 609–623.
9 GARTLING, D. K., Convective heat transfer analysis by the finite element method, Comp. Meth. Appl. Mech. Engng. **12** (1977), S. 365–382.
10 GARTLING, D. K., Some comments on [11], Int. J. Num. Meth. Engng. **12** (1978), S. 187–190.
11 HEINRICH, J. C.; HUYAKORN, P. S.; ZIENKIEWICZ, O. C.; MITCHELL, A. R., An 'upwind' finite element scheme for two-dimensional convective transport equation, Int. J. Num. Meth. Engng. **11** (1977), S. 131–143.
12 HOLST, P. H.; AZIZ, K., Transient three-dimensional natural convection in confined porous media, Int. J. Heat & Mass Transfer **15** (1972), S. 73–90
13 HOOD, P.; TAYLOR, C., NAVIER-STOKES equations using mixed interpolation, in: ODEN, J. T.; ZIENKIEWICZ, O. C.; GALLAGHER, R. H.; TAYLOR, C. (Ed.), Finite Element Methods in flow problems, Proc. Int. Symp., Swansea 7.–11. 1. 74, Huntsville, Alabama: UAH-Press 1974, S. 121–132.
14 HOOD, P., Frontal solution program for unsymmetric matrices, Int. J. Num. Meth. Engng. **10** (1976), S. 379–399.
15 HORNE, R. N.; O'SULLIVAN, M. J., Oscillatory convection in a porous medium heated from below, J. Fluid Mech. **66** part 2 (1974), S. 339–352.
16 HUYAKORN, P. S.; TAYLOR, C., Finite element models for coupled groundwater flow and convective dispersion, in: GRAY, W. G.; PINDER, G. F.; BREBBIA, C. A. (Ed.), Finite elements in water resources, Proc. 1st Int. Conf. Princeton Univ. July 1976, London-Plymouth: Pentech Press 1977, S. 1.131–1.151.
17 KVERNVOLD, O.; TYVAND, P. A., Dispersion effects on thermal convection in porous media, J. Fluid Mech. **99** part 4 (1980), S. 673–686.
18 MEISSNER, U., A mixed finite element model for use in potential flow problems, Int. J. Num. Meth. Engng. **6** (1973), S. 467–473.
19 NIELD, D. A., Onset of thermohaline convection in a porous medium, Water Resources Research **4** (1968), S. 553–560.
20 PALM, E.; WEBER, J. E.; KVERNVOLD, O., On steady convection in a porous medium, J. Fluid Mech. **54** part 1 (1972), S. 153–161.
21 RUBIN, H., Effect of solute dispersion on thermal convection in a porous medium layer, 2, Water Resources Research **11** (1975), S. 154–158.
22 RUBIN, H.; CHRISTENSEN, B. A., Convection currents associated with hydrodynamic dispersion in a porous medium, Proc. IAHR 16th Congr. Sao Paulo, Brazil, 27. 7.–1. 8. 75. **5** (1975), S. 162–168.
23 RUBIN, H., On the analysis of cellular convection in porous media, Int. J. Heat & Mass Transfer **18** (1975), S. 1483–1486.
24 SCHUBERT, G.; STRAUS, J. M., Three-dimensional and multicellular steady and unsteady convection in fluid-saturated porous media at high RAYLEIGH numbers, J. Fluid Mech. **94** part 1 (1979), S. 25–38.
25 SEGOL, G.; PINDER, G. F.; GRAY, W. G., A GALERKIN-finite element technique for calculating the transient position of the salt-water front, Water Resources Research **11** (1975), S. 343–347.
26 STRAUS, J. M., Large amplitude convection in porous media, J. Fluid Mech. **64** part 1 (1974), S. 51–63.
27 TABARROK, B.; LIN, R. C., Finite element analysis of free convection flows, Int. J. Heat & Mass Transfer **20** (1977), S. 945–952.
28 TAYLOR, C.; HUYAKORN, P. S., Three-dimensional groundwater flow with convective dispersion, in: GALLAGHER, R. H.; ZIENKIEWICZ, O. C.; ODEN, J. T.; CECCHI, M. M.; TAYLOR, C. (Ed.), Finite elements in fluids **3** (1978), J. Wiley & Sons, Chichester/New York/Brisbane/Toronto, S. 311–321.
29 VAN GENUCHTEN, M. TH.; GRAY, W. G., Analysis of some dispersion corrected numerical schemes for solution of the transport equation, Int. J. Num. Meth. Engng. **12** (1978), S. 387–404.
30 WEBER, J. E., Dispersion effect on buoyancy-driven convection in stratified flows through porous media, J. of Hydrology **25** (1975), S. 59–70.
31 ZIENKIEWICZ, O. C., The finite element method, 3rd Edition, McGraw Hill, London et al. 1977.
32 ZIENKIEWICZ, O. C.; HEINRICH, J. C., The finite element method and convection problems in fluid mechanics, in: GALLAGHER, R. H.; ZIENKIEWICZ, O. C.; ODEN, J. T.; CECCHI, M. M.; TAYLOR, C. (Ed.), Finite elements in fluids **3** (1978), J. Wiley & Sons, Chichester/New York/Brisbane/Toronto, S. 1–22.

Eingereicht am 4. 7. 1980, in revidierter Fassung am 5. 1. 1981

Anschrift: Dr. HANS-JÖRG DIERSCH, Institut für Mechanik der AdW der DDR, DDR-1199 Berlin, Rudower Chaussee 5

KLEINE MITTEILUNGEN

ZAMM 61, 339 – 341 (1981)

A. Chakrabarti

On a System of Wiener-Hopf Equations Associated with a Mixed Diffraction Problem.

Correct solutions of a system of Wiener-Hopf equations derived in an earlier paper [1] have been obtained by reducing the system to an equivalent system of singular integral equations. The analysis of the present work is of general application and modifies the Wiener-Hopf technique in a proper manner.

Introduction

The following system of Wiener-Hopf equations

$$2i(s^2+k^2)^{1/2}\,V_-(s,0)+i(s^2+k^2)^{1/2}\,[V_+(s,0^+)+V_+(s,0^-)]= \\ =P_+(s,0^-)-P_+(s,0^+) \tag{1}$$

and

$$-i(s^2+k^2)^{1/2}[V_+(s,0^+)-V_+(s,0^-)]= \\ =2P_-(s,0)+P_+(s,0^+)+P_+(s,0^-) \tag{2}$$

arose in [1], in connection with a mixed diffraction problem, involving the solution of the reduced wave equation

$$(\nabla^2+k^2)\,v=0\,,\qquad -\infty<x,y<\infty\,, \tag{3}$$

under the mixed boundary conditions

$$v(x,0^+)=-u_0(x,0)\,,\qquad \frac{\partial v(x,0^-)}{\partial y}=-\frac{\partial u_0(x,0)}{\partial y}\,,\qquad x>0\,, \tag{4}$$

the continuity conditions

$$v(x,0^+)=v(x,0^-)\,,\qquad \frac{\partial v(x,0^+)}{\partial y}=\frac{\partial v(x,0^-)}{\partial y}\,,\qquad x<0\,, \tag{5}$$

a radiation condition at infinity along with certain edge conditions at the edge $x=0^+$, to ensure uniqueness of the solution.

In equations (1) and (2), the functions V and P are defined as

$$\begin{aligned} V_+(s,0^\pm)&=\lim_{y\to 0\pm}\int_0^\infty v(x,y)\,e^{-sx}\,dx\,,\\ V_-(s,0)\equiv V_-(s,0^\pm)&=\lim_{y\to 0\pm}\int_{-\infty}^0 v(x,y)\,e^{-sx}\,dx\,,\\ P_+(s,0^\pm)&=\lim_{y\to 0\pm}\int_0^\infty \frac{\partial v}{\partial y}\,e^{-sx}\,dx\,,\\ P_-(s,0)\equiv P_-(s,0^\pm)&=\lim_{y\to 0\pm}\int_{-\infty}^0 \frac{\partial v}{\partial y}\,e^{-sx}\,dx\,. \end{aligned} \tag{6}$$

The function u_0 is assumed to be of such a type that under a further assumption that $k=k_r-ik_i$ $(k_i>0)$ (k is actually real and this assumption is essential to employ the Wiener-Hopf technique), all the Laplace transforms defined in equations (6) exist and are analytic in a strip: $\sigma_-<\sigma<\sigma_+$ of the complex $s\ (=\sigma+i\tau)$-plane, which is contained in the strip $-k_i<\sigma<+k_i$. The functions with a suffix $(+)$ are analytic in the right half-plane $\sigma>\sigma_-$ and those with a suffix $(-)$ are analytic in the left half-plane $\sigma<\sigma_+$.

The edge conditions that were assumed in [1], as well as in [2], have the following forms

$$v(x,0^+)\sim O(x^{1/4})\,,\qquad \frac{\partial v(x,0^+)}{\partial y}\sim (Ox^{-(3/4)})\,, \tag{7}$$

as $x\to 0^+$, with usual meaning of the 'O' symbols.

In the system of equations (1) and (2), there are altogether four unknown functions $V_-(s,0)$, $P_-(s,0)$, $V_+(s,0^-)$ and $P_+(s,0^+)$. The functions $V_+(s,0^+)$ and $P_+(s,0^-)$ are known through the mixed boundary conditions (4).

In trying to solve a system of Wiener-Hopf equations of the type of equations (1) and (2), Rawlins [2] and Hurd [3] have modified the usual Wiener-Hopf technique in some particular fashions which require the validity of the original Wiener-Hopf equations even outside the strip $\sigma_-<\sigma<\sigma_+$ in which these equations are actually valid.

The present author had pointed out in [1] this defect of modification of the Wiener-Hopf technique proposed in [2] and had reduced the system of equations (1) and (2) into the following system of integral equations, by employing the usual Wiener-Hopf arguments:

$$\begin{aligned} &\frac{1}{2\pi i}\int_{a-i\infty}^{a+i\infty}\frac{dw}{w-s}[i(w-ik)^{1/2}\,V_+(w,0^-)+\\ &+(w+ik)^{-(1/2)}\,P_+(w,0^+)]=Q_1(s),\\ &\frac{1}{2\pi i}\int_{a-i\infty}^{a+i\infty}\frac{dw}{w-s}[-i(w+ik)^{1/2}\,V_+(w,0^-)+\\ &+(w-ik)^{-(1/2)}\,P_+(w,0^+)]=Q_2(s)\,,\\ &\mathrm{Re}(s)>a\,,\qquad \sigma_-<a<\sigma_+\,, \end{aligned} \tag{8}$$

where the known functions Q_1 and Q_2 are given by

$$\begin{aligned} Q_1(s)&=\frac{1}{2\pi i}\int_{a-i\infty}^{a+i\infty}\frac{dw}{w-s}[(w+ik)^{-(1/2)}\,P_+(w,0^-)-\\ &\quad -i\,(w-ik)^{1/2}\,V_+(w,0^+)]\\ &\text{and}\\ Q_2(s)&=\frac{1}{2\pi i}\int_{a-i\infty}^{a+i\infty}\frac{dw}{w-s}[-i(w+ik)^{1/2}\,V_+(w,0^+)-\\ &\quad-(w-ik)^{-(1/2)}\,P_+(w,0^-)]\,. \end{aligned} \tag{9}$$

Without being able to discover a method of solving the integral equations (8) with general form of the right hand sides, that accommodate any incident wave, a particular case was considered in [1], which corresponded to an incident plane wave and for which $u_0(x,y)$ was taken in the form:

$$u_0(x,y)=\exp\left[-ik(x\cos\varphi_0+y\sin\varphi_0)\right]\,,\qquad 0<\varphi_0<\pi\,. \tag{10}$$

The particular integral equations of the type (8) were then solved in [1] by borrowing the idea of modification of the Wiener-Hopf technique due to Rawlins [2] and it was found that the solution of the Wiener-Hopf equations (1) and (2) agreed, in principle, with the solution of a similar problem considered in [2].

The defect in Rawlins's modification of the Wiener-Hopf technique was thus merely left as pointed out in [1], but not actually remedied there.

In the present paper we describe a procedure of solving the Wiener-Hopf system (1) and (2) fot a sufficiently general u_0, which is free from the defects mentioned in [1] and obtain completely different forms of the solution in the particular case when u_0 is given by (10).

The analysis of the present work is of general application and may be regarded as a proper modification of the usual Wiener-Hopf technique described in [4] and [5]. A similar analysis may be employed in the problem considered by Hurd, too.

The mathematical problem of solving the reduced wave equation (3) under mixed boundary conditions of the type (4) has also been considered by Williams [6] and by Chakrabarti and Sastry [7] by means of analyses avoiding the Wiener-Hopf technique.

We observe that the edge conditions (7) are not the correct ones for the diffraction problem considered in [1] and [2]. It is, in fact, difficult to deduce the appropriate edge conditions, and we propose here the modified edge conditions through equations (11), in order to obtain the unique solution of the system of WIENER-HOPF equations (1) and (2).

2. Method of solution

Introduce a branch cut L emanating from the point $w = ik$ and going towards infinity, in the right half of the w-plane, in the direction making an angle arg $(s - ik)$ with the real axis, so that the point $w = s$ lies on the cut L.

Close the contours for the integrals on the left hand sides of both the equations (8) by means of semi-circular arcs in the right half-plane along with a loop around the cut L.

We then propose the following edge conditions for the mixed diffraction problem considered in [1]:

$$v(x, 0\pm) \sim O(1)\,, \quad \frac{\partial v(x, 0\pm)}{\partial y} \sim O\left(x^{-(1/2)} \log x\right) \quad \text{as} \quad x \to 0^+\,. \tag{11}$$

(Edge conditions in the form (7) are actually difficult to obtain by employing KARP and KARAL's analysis [8].) and use the ABELian theorems (see [5]) for LAPLACE transforms.

We thus find that the integral equations (8) can be recast into the following forms:

$$-(s + ik)^{-(1/2)}\, P_+(s, 0^+) + \frac{1}{\pi i} \int\limits_L \frac{i(\omega - ik)^{1/2} V_+(w, 0^-)}{w - s}\, dw = Q_1(s)\,, \tag{12}$$

$$(s + ik)^{1/2}\, i V_+(s, 0^-) + \frac{1}{\pi i} \int\limits_L \frac{(w - ik)^{-(1/2)}\, P_+(w, 0^+)}{w - s}\, dw = Q_2(s)\,, \tag{13}$$

for $s \in L$.

In deriving the equations (12) and (13) from equations (8), by the procedure mentioned above, it has been made use of the fact that the function $(w - ik)^{1/2}$ takes up the value $-(w - ik)^{1/2}$ when the point w changes its position from the immediate right of the cut L to the immediate left of it while moving along the loop around L.

It is to be noted here that the same integral equations (12) and (13) can also be derived directly from the original WIENER-HOPF equations (1) and (2), after rewriting them in a suitable manner, using the factorisation: $(s^2 + k^2)^{1/2} = (s + ik)^{1/2} \cdot (s - ik)^{1/2}$, multiplying both sides by $\frac{1}{2\pi i}\, \frac{dw}{w - s}$ $[\text{Re}(s) > a]$ and integrating along the line $\text{Re}(w) = a$, lying completely within the strip of validity of the equations and making use of the edge conditions (11) along with CAUCHY's residue theorem, after closing the contour for each integral in the appropriate half-plane. This latter procedure of deriving the integral equations (12) and (13) from the original WIENER-HOPF equations (1) and (2) may be regarded as the real modification of the usual WIENER-HOPF technique described in [4]. We note that the present method of reduction of a system of WIENER-HOPF equations to a system of singular integral equations is free from the defect that is inherent in the modification of the WIENER-HOPF technique proposed in [2] and [3].

The integral equations (12) and (13) are special cases of a system of singular integral equations, studied in detail by PETERS [9], and can be solved by utilizing the usual RIEMANN-HILBERT method described in MUSKHELISHVILI's book [10]. (Note that HURD [3] also arrived at a system of HILBERT problems for the system of WIENER-HOPF equations handled by him and had called the method of solution as the "WIENER-HOPF-HILBERT method".)

Solving the associated RIEMANN-HILBERT problems for the system of singular integral equations (12) and (13), by the method described in [10], we find that

$$P_+(s, 0^+)\, (s + ik)^{-(1/2)} = -\frac{1}{2}\, Q_1(s) + \frac{1}{2\pi i} \int\limits_L \left\{\frac{w - ik}{w + ik}\right\}^{1/2} \frac{Q_2(w)}{w - s}\, dw\,, \tag{14}$$

$$i V_+(s, 0^-) = \frac{1}{2}\, (s + ik)^{-(1/2)}\, Q_2(s) + \frac{1}{2\pi i} \int\limits_L \frac{Q_1(w)\, dw}{(w - ik)^{1/2} \cdot w - s}\,, \tag{14}$$

for $s \in L$.

The equations (14) represent the actual solution of the system of WIENER-HOPF equations (1) and (2), if the principle of analytic continuation is employed.

3. A particular case

If, in particular (see [1]), we take u_0 to be given by equation (10), we find that

$$Q_1(s) = -\frac{1}{s + ik \cos \varphi_0} \left[i(-2ik)^{1/2} \cos \frac{\varphi_0}{2} + ik \sin \varphi_0 \cdot (s + ik)^{-(1/2)} \right], \tag{15}$$

$$Q_2(s) = -\frac{1}{s + ik \cos \varphi_0} \left[i(s + ik)^{1/2} + (-2ik)^{1/2} \sin \frac{\varphi_0}{2} \right], \tag{16}$$

and that the solutions (14) take the forms:

$$\begin{aligned} P_+(s, 0^+)\, (s + ik)^{-(1/2)} &= -\frac{1}{2}\, Q_1(s) - \frac{(-2ik)^{1/2}}{2\pi i(s + ik \cos \varphi_0)} \cdot \Bigg[-\pi \cos \frac{\varphi_0}{2} + \sin \frac{\varphi_0}{2} \times \\ &\quad \times \{(s - ik)\, J_1(s) + ik(1 + \cos \varphi_0)\, J_1(-ik \cos \varphi_0)\} \Bigg], \\ i V_+(s, 0^-) &= \frac{1}{2}\, (s + ik)^{-(1/2)}\, Q_2(s) - \frac{1}{2\pi i(s + ik \cos \varphi_0)} \cdot [-\pi + ik \sin \varphi_0 \times \\ &\quad \times \{J_1(s) - J_1(-ik \cos \varphi_0)\}]\,, \end{aligned} \tag{17}$$

where the function J_1 is given as in equations (18), quoted later.

In deriving equations (17) we have used the following results, obtained by standard methods (see [4], for example):

$$\begin{aligned} J_1(\nu) &\equiv \int\limits_L \frac{dw}{(w^2 + k^2)^{1/2}\, (w - \nu)} = \frac{2i}{(\nu^2 + k^2)^{1/2}} \cos^{-1} (\nu/ik) + \frac{\pi i}{(\nu^2 + k^2)^{1/2}}\,, \\ &\quad \nu \notin L, \text{ but belongs to the half-plane } \sigma > a, \\ J_1(s) &= \frac{2i}{(s^2 + k^2)^{1/2}} \cos^{-1} (s/ik)\,, \qquad s \in L\,, \\ J_2(\nu) &\equiv \frac{1}{\pi i} \int\limits_L \frac{dw}{(w - ik)^{1/2}\, (w - \nu)} = \begin{cases} \dfrac{1}{(\nu - ik)^{1/2}}\,, & \nu \notin L\,, \\ 0, & \nu \in L\,. \end{cases} \end{aligned} \tag{18}$$

We observe that the function $J_1(s)$ is analytic in the right half-plane $\sigma > a$ (see NOBLE [4]), and the expressions obtained in equations (17) for the functions $V_+(s, 0^-)$ and $P_+(s, 0^+)$ actually represent two right-analytic functions as expected.

We also observe that

$$V_+(s, 0^-) \sim O(s^{-1})\,, \qquad P_+(s, 0^+) \sim O(s^{-(1/2)} \log s)\,, \tag{19}$$

as opposed to the conditions

$$V_+(s, 0^-) \sim O(s^{-(5/4)})\,, \qquad P_+(s, 0^+) \sim O(s^{-(1/4)})\,, \tag{20}$$

as $|s| \to \infty$ in the right half-plane, the latter results (20) (see [1] and [2]) being due to the assumption of the edge conditions in the form given by equations (7). Equations (19) follow from the edge conditions proposed by means of equations (11) (see [5]).

4. Conclusion

The question raised in [1] about the modification of the WIENER-HOPF technique proposed in [2] is now answered completely. The analysis of the present work is expected to be successfully applicable to a class of problems for which the system of singular integral equations of the type (12) and (13) can be solved completely (see [9]). The present modification of the WIENER-HOPF technique essentially requires the factorisation of a single function (or a set of functions) ($[s^2 + k^2]^{1/2} = [s + ik]^{1/2} [s - ik]^{1/2}$, in the present problem), instead of a matrix of functions, analytic in a strip.

References

1 CHAKRABARTI, A., A direct approach to the problem of diffraction by a half-plane under mixed boundary conditions, ZAMM **59**, 241–246 (1979).
2 RAWLINS, A. D., The solution of a mixed boundary value problem in the theory of diffraction by a semi-infinite plane, Proc. Roy. Soc. London, Ser. A **346**, 469 (1975).
3 HURD, R. A., The WIENER-HOPF-HILBERT method for diffraction problems, Canadian J. Phys. **54**, 775 (1976).
4 NOBLE, B., The WIENER-HOPF Technique, London, Pergamon 1958.
5 JONES, D. S., The Theory of Electromagnetism, London, Pergamon 1964.
6 WILLIAMS, W. E., A note on GREEN's functions for the HELMHOLTZ equation Quart. J. Mech. Appl. Math. **31**, 261 (1978).
7 CHAKRABARTI, A.; SASTRY, V. V. S. S., Diffraction of a cylindrical pulse by a half-plane under mixed boundary conditions, Canadian J. Phys. **57**, 1324 (1979).
8 KARP, S. N.; KARAL, F. C., Commun. pure appl. Math. **15**, 413 (1962).
9 PETERS, A. S., Commun. pure appl. Math. **25**, 369 (1972).
10 MUSKHELISHVILI, N. I., Singular Integral Equations, Noordhoff, Groningen, Holland 1953.

Eingereicht am 28. 8. 1980

Anschrift: Dr. A. CHAKRABARTI, Department of Applied Mathematics, Indian Institute of Science, Bangalore-560012, INDIA

ZAMM 61, 341–342 (1981)

A. A. RAPTIS / C. P. PERDIKIS / G. J. TJIVANIDIS

Effects of Free Convection Currents on the Flow of an Electrically Conducting Fluid past an Accelerated Vertical Infinite Plate with Variable Suction

1. Introduction

The problem of free convection of an electrically conducting fluid past a porous plate under the influence of a magnetic field has attracted the interest of many authors in view of its applications to geophysics, astrophysics, engineering and to the boundary layer control in the field of aerodynamics.

SOUNDALGEKAR [1] has derived an exact solution for the flow past an impulsively started infinite vertical plate in its own plane, known as STOKES' problem for the vertical plate. Recently POP and SOUNDALGEKAR [2] have studied the free convection flow past an accelerated vertical infinite plate.

The purpose of this paper is to study the unsteady hydromagnetic free convection flow past a non-conducting infinite vertical porous plate in the presence of a transverse magnetic field. The magnetic REYNOLDS number of the flow is taken to be small enough so that the induced magnetic field is negligible.

2. Mathematical Analysis

All the geometry and the assumptions are described in introduction and ref. of [2]. The equations governing the flow are:

$$\frac{\partial v^+}{\partial y^+} = 0\,, \tag{1}$$

$$\frac{\partial u^+}{\partial t^+} + v^+ \frac{\partial u^+}{\partial y^+} = g\beta(T^+ - T^+_\infty) + \nu \frac{\partial^2 u^+}{\partial y^{+2}} - \frac{\sigma B_0^2}{\varrho} u^+\,, \tag{2}$$

$$\frac{\partial T^+}{\partial t^+} + v^+ \frac{\partial T^+}{\partial y^+} = \frac{k}{\varrho c_p} \frac{\partial^2 T^+}{\partial y^{+2}}\,, \tag{3}$$

where v^+ is the velocity component perpendicular to the plate, σ the electrical conductivity of the fluid, B_0 the magnetic induction and all other physical variables have already been defined in [2]. In equation (3) the JOULE heating term is neglected because it is of the same order of magnitude with the viscous dissipation term.

The initial and boundary conditions corresponding to the present problem are:

$$\begin{aligned} &t^+ \leqq 0: u^+ = 0\,, \quad T^+ = T^+_\infty \quad \text{everywhere}\,,\\ &\qquad u^+ = U^+(t^+)\,, \quad T^+ = T^+_w \quad \text{at} \quad y^+ = 0\,,\\ &t^+ > 0:\\ &\qquad u^+ \to 0\,, \quad T^+ \to T^+_\infty \quad \text{as} \quad y^+ \to \infty\,. \end{aligned} \tag{4}$$

From equation (1), we take

$$v^+ = -v_0^+(t^+) = -a\left(\frac{\nu}{t^+}\right)^{1/2},$$

where a represents the velocity of suction or injection at the plate according as $a > 0$ or $a < 0$ respectively.

If we now introduce the following dimensionless quantities

$$\begin{aligned} &y = \frac{U_0^+ y^+}{\nu}\,, \quad t = \frac{U_0^{+2} t^+}{\nu}\,, \quad u = \frac{u^+}{U_0^+}\,, \quad U = \frac{U^+}{U_0^+}\,,\\ &T = \frac{(T^+ - T^+_\infty)}{(T^+_w - T^+_\infty)}\,, \quad Gr = \frac{g\beta\nu(T^+_w - T^+_\infty)}{U_0^{+3}}\,,\\ &Pr = \frac{\mu c_p}{k}\,, \quad M = \frac{\sigma B_0^2 \nu}{\varrho U_0^{+2}} \end{aligned} \tag{5}$$

where Gr, Pr and M denote the GRASHOF, PRANDTL and magnetic field parameters respectively, then Eqs. (2) and (3) become

$$\frac{\partial u}{\partial t} - \frac{a}{t^{1/2}} \frac{\partial u}{\partial y} = Gr\,T + \frac{\partial^2 u}{\partial y^2} - Mu\,, \tag{6}$$

$$\frac{\partial T}{\partial t} - \frac{a}{t^{1/2}} \frac{\partial T}{\partial y} = \frac{1}{Pr} \frac{\partial^2 T}{\partial y^2}\,. \tag{7}$$

Using (4) and (5), the corresponding initial and boundary conditions in dimensionless form are

$$\begin{aligned} &t \leqq 0: \quad u = 0\,, \quad T = 0 \quad \text{everywhere}\,,\\ &\qquad u = U(t)\,, \quad T = 1 \quad \text{at} \quad y = 0\,,\\ &t > 0:\\ &\qquad u \to 0\,, \quad T \to 0 \quad \text{as} \quad y \to \infty\,. \end{aligned} \tag{8}$$

Assuming the magnetic field parameter M to be small, we expand the variable u as follows:

$$u(y, t) = u_0(y, t) + M u_1(y, t) + \dots \tag{9}$$

By substituting (9) into (6) and (7) and equating the coefficients of the same powers of M we get the following hierarchy of equations when $Pr = 1$. (The PRANDTL number Pr is a measure of the relative importance of viscosity and heat conductivity in the fluid. For most gases Pr is of unit order, so that the velocity and the thermal boundary layers will be of the same order of thickness and we will seek a solution of our problem for this case).

$$\frac{\partial u_0}{\partial t} - \frac{a}{t^{1/2}} \frac{\partial u_0}{\partial y} = Gr\,T + \frac{\partial^2 u_0}{\partial y^2}\,, \tag{10}$$

$$\frac{\partial u_1}{\partial t} - \frac{a}{t^{1/2}} \frac{\partial u_1}{\partial y} = \frac{\partial^2 u_1}{\partial y^2} - u_0\,, \tag{11}$$

$$\frac{\partial T}{\partial t} - \frac{a}{t^{1/2}} \frac{\partial T}{\partial y} = \frac{\partial^2 T}{\partial y^2}\,. \tag{12}$$

The initial and boundary conditions for an uniformly accelerated plate $U(t) = t$ are obtained from (8)

$$t \leqq 0: \quad u_0(y, t) = u_1(y, t) = 0 , \qquad T(y, t) = 0 \quad \text{everywhere} ,$$

$$u_0(0, t) = t , \qquad u_1(0, t) = 0 , \qquad T(0, t) = 1 , \tag{13}$$

$$t > 0:$$

$$u_0(\infty, t) \to 0 , \qquad u_1(\infty, t) \to 0 , \qquad T(\infty, t) \to 0 .$$

A solution of equations (10) and (11) for small t can be obtained in the form

$$u_0 = t f_0(\eta) , \qquad u_1 = t^2 f_1(\eta) , \ldots \tag{14}$$

where $\eta = \dfrac{y}{2t^{1/2}}$.

Introducing (14) into Eqs (10)—(12) we obtain

$$T'' + 2(\eta + a)\, T' = 0 , \tag{15}$$

$$f_0'' + 2(\eta + a) f_0' - 4 f_0 = -4 Gr T , \tag{16}$$

$$f_1'' + 2(\eta + a) f_1' - 8 f_1 = 4 f_0 , \tag{17}$$

where a prime denotes differentiation with respect to η. The corresponding boundary conditions become

$$\begin{aligned} f_0(0) &= 1 , \quad f_1(0) = 0 , \quad T(0) = 1 , \\ f_0(\infty) &= 0 , \quad f_1(\infty) = 0 , \quad T(\infty) = 0 . \end{aligned} \tag{18}$$

Thus the solutions of (15)—(17) by virtue of the boundary conditions (18) are:

$$T(\eta) = \frac{Hh_0(\sqrt{2}(\eta + a))}{Hh_0(\sqrt{2}\, a)} , \tag{19}$$

$$f_0(\eta) = (1 - Gr) \frac{Hh_2(\sqrt{2}\, (\eta + a))}{Hh_2(\sqrt{2}\, a)} + Gr \frac{Hh_0(\sqrt{2}\, (\eta + a))}{Hh_0(\sqrt{2}\, a)} , \tag{20}$$

$$f_1(\eta) = \frac{2 - Gr}{2} \frac{Hh_4(\sqrt{2}\, (\eta + a))}{Hh_4(\sqrt{2}\, a)} + (Gr - 1) \frac{Hh_2(\sqrt{2}\, (\eta + a))}{Hh_2(\sqrt{2}\, a)} - \frac{Gr}{2} \frac{Hh_0(\sqrt{2}\, (\eta + a))}{Hh_0(\sqrt{2}\, a)} . \tag{21}$$

The function Hh_n (η: ... —1, 0, 1) is defined in JEFFREYS and JEFFREYS [3].

Using the expressions (20) and (21), we can calculate the skin friction τ_w. This expression is given by the relation

$$\tau_w = \frac{1}{2} t^{1/2} [f_0'(0) + M t f_1'(0)] \tag{22}$$

where

$$f_0'(0) = -\sqrt{2} \left[(1 - Gr) \frac{Hh_1(\sqrt{2}\, a)}{Hh_2(\sqrt{2}\, a)} + Gr \frac{Hh_{-1}(\sqrt{2}\, a)}{Hh_0(\sqrt{2}\, a)} \right] , \tag{23}$$

$$f_1'(0) = -\sqrt{2} \left[\frac{(2 - Gr)}{2} \frac{Hh_3(\sqrt{2}\, a)}{Hh_4(\sqrt{2}\, a)} + (Gr - 1) \frac{Hh_1(\sqrt{2} a)}{Hh_2(\sqrt{2}\, a)} - \frac{Gr}{2} \frac{Hh_{-1}(\sqrt{2}\, a)}{Hh_0(\sqrt{2}\, a)} \right] . \tag{24}$$

3. Results

The velocity profiles are shown in Fig. 1 and 2 for different values of the magnetic parameter M and the parameter a, when GRASHOF number $Gr > 0$ and $Gr < 0$, respectively. Physically, the $Gr < 0$ corresponds to an externally heated plate as the free-convection currents are carried towards the plate while $Gr > 0$ to an externally cooled plate. The parameter a represents the velocity of suction or injection at the plate when $a > 0$ or $a < 0$ respectively.

From Fig. 1 we observe that application of suction ($a = 0.5$) helps in decreasing of the velocity which is greater for higher values of magnetic parameter M. In the case $Gr < 0$ there is a reverse type of flow (Fig. 2) and this phenomenon becomes weak for higher values of the magnetic parameter M.

The numerical values of skin-friction τ_w are entered in the accompanied table for some values of Gr, M and a. We conclude from this table that the skin-friction decreases with increasing the magnetic parameter or the parameter a.

Table I. Values of Skin-friction (τ_w)

$Pr = 1$, $t = 0.2$

M	a	Gr	
		2	−2
0.0	−0.5	0.135991	−0.924915
0.0	0.5	−0.181559	−1.094542
0.3	−0.5	0.120078	−0.929368
0.3	0.5	−0.195254	−1.100233

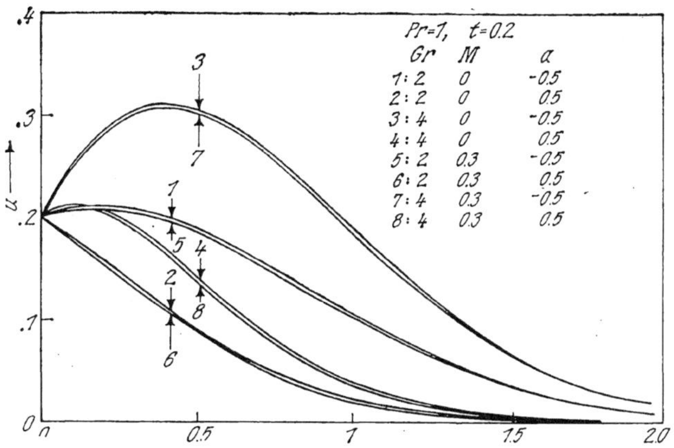

Fig. 1. Velocity profiles for $Gr > 0$

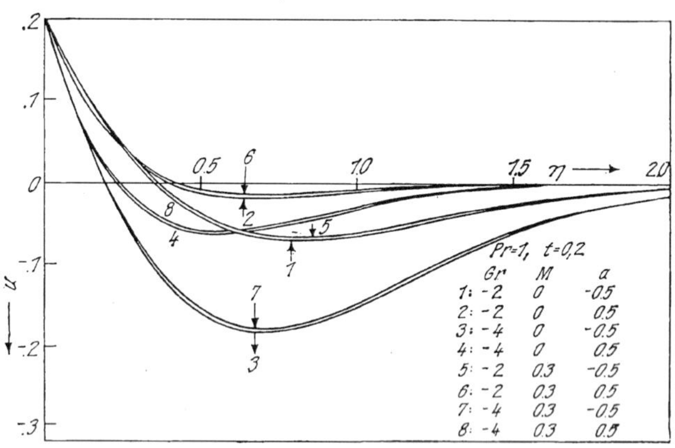

Fig. 2. Velocity profiles for $Gr < 0$

References

1 SOUNDALGEKAR, V. M., Free convection effects on the STOKES problem for an infinite vertical plate, Trans. ASME, J. Heat Transfer 99, 499—501 (1977).

2 POP, I. and SOUNDALGEKAR, V. M., Free convection flow past an accelerated vertical infinite plate, ZAMM 60, 167—168 (1980).

3 JEFFREYS, H. and JEFFREYS, B. S., Methods of Mathematical Physics, Cambridge University Press, p. 622 (1972).

Eingereicht am 2. 11. 1980

Anschrift: Dr. A. RAPTIS, Dr. C. PERDIKIS, Prof. G. TZIVANIDIS, University Ioannina, School of Physics and Mathematics, Ioannina, Greece

ZAMM 61, 343 – 344 (1981)

L. Jentsch

Der Greensche Kontakttensor der Elastostatik für zwei fest verbundene Halbebenen

Der Greensche Kontakttensor der Elastostatik für zwei längs der Ebene $x_3 = 0$ fest verbundene Halbräume wurde in [1] explizit ermittelt. Er enthält 11 Konstanten, die einem linearen Gleichungssystem genügen (s. [1], (7.24)–(7.27), (7.29)–(7.35)). Später wurde dieses Gleichungssystem gelöst und die Konstanten sind in [2] angegeben.

Der Kontakttensor für zwei verbundene Halbebenen wurde durch Integration der einzelnen Glieder der zweiten und dritten Spalte des räumlichen Kontakttensors bezüglich y_1 von minus bis plus unendlich gefunden. Es wird also über die an der Stelle $y = (y_1, y_2, y_3)$ in y_2- und y_3-Richtung wirkende Einzelkraft bezüglich y_1 summiert. Dabei ist bei auftretenden divergenten Integralen der endliche Teil zu nehmen, genauer: wir bilden

$$\lim_{l\to\infty} \left[\int_{+l}^{-l} \dots \mathrm{d}y_1 - 2\ln l \right] \text{ bei divergenten Integralen},$$

$$\lim_{l\to\infty} \int_{-l}^{+l} \dots \mathrm{d}y_1 \text{ bei konvergenten Integralen}.$$

Man erhält bis auf den Faktor 2 folgendes Resultat (die Variablen Nr. 2 und 3 sind ab jetzt mit Nr. 1 und 2 bezeichnet, $x = (x_1, x_2)$, $y = (y_1, y_2)$):

$$G(x, y) = \begin{cases} \underset{(1)}{\Gamma}(x, y) + V(x, y) & \text{für } x_2 > 0, y_2 > 0 \\ \underset{(0)}{\Gamma}(x, y) + V(x, y) & \text{für } x_2 < 0, y_2 < 0 \\ V(x, y) & \text{für } x_2 > 0, y_2 < 0 \text{ und} \\ & \phantom{\text{für }} x_2 < 0, y_2 > 0. \end{cases}$$

Hierbei ist $\underset{(i)}{\Gamma}(x, y)$ die zweireihige Somiglianasche Grundlösungsmatrix für den Operator der ebenen Elastostatik $\underset{(i)}{A}(\partial_x)$ mit den Komponenten $\underset{(i)}{A_{jk}}(\partial_x) = \mu_i \delta_{jk} \Delta + (\lambda_i + \mu_i) \dfrac{\partial^2}{\partial x_j\, \partial x_k}$, $j, k = 1, 2$ (s. [3]). Die Elemente von $\underset{(i)}{\Gamma}(x, y)$ sind

$$\underset{(i)}{\Gamma_{jk}}(x, y) = \frac{\lambda_i + 3\mu_i}{2\mu_i(\lambda_i + 2\mu_i)} \delta_{jk} \ln \frac{1}{|x - y|} + \frac{\lambda_i + \mu_i}{2\mu_i(\lambda_i + 2\mu_i)} \frac{(x_j - y_j)(x_k - y_k)}{|x - y|^2}.$$

Die Kompensatrix $V(x, y)$ kann mit Hilfe der Potentialfunktionen

$$Z(x_1, x_2; y_1, y_2) = \ln \sqrt{(x_1 - y_1)^2 + (|x_2| + |y_2|)^2},$$

$$W(x_1, x_2; y_1, y_2) = \arctan \frac{x_1 - y_1}{|x_2| + |y_2|}$$

(Z und W sind für $x_2 > 0$ und Z und $-W$ für $x_2 < 0$ zueinander konjugierte Potentialfunktionen) wie folgt geschrieben werden: Für $x_2 > 0$, $y_2 > 0$ ist

$$V(x, y) = -H_6^+ \begin{pmatrix} 1 & 0 \\ 0 & 1 \end{pmatrix} Z + \begin{pmatrix} -H_7^+ & -H_8^+ W \\ H_8^+ W & 0 \end{pmatrix} + \begin{pmatrix} -(H_9^+ x_2 + H_{10}^+ y_2) \dfrac{\partial Z}{\partial x_2} & (H_9^+ x_2 - H_{10}^+ y_2) \dfrac{\partial Z}{\partial x_1} \\ (H_9^+ x_2 - H_{10}^+ y_2) \dfrac{\partial Z}{\partial x_1} & (H_9^+ x_2 + H_{10}^+ y_2) \dfrac{\partial Z}{\partial x_2} \end{pmatrix} + H_{11}^+ x_2 y_2 \begin{pmatrix} \dfrac{\partial^2 Z}{\partial x_2^2} & \dfrac{\partial^2 Z}{\partial x_1\, \partial x_2} \\ -\dfrac{\partial^2 Z}{\partial x_1\, \partial x_2} & \dfrac{\partial^2 Z}{\partial x_2^2} \end{pmatrix}. \tag{1}$$

Für $x_2 < 0$, $y_2 < 0$ ist in (1) H_k^+ durch H_k^- ($k = 6, 7, 9, 10, 11$) und H_8^+ durch $-H_8^-$ zu ersetzen.

Für $x_2 < 0$, $y_2 > 0$ ist

$$V(x, y) = -H_1^+ \begin{pmatrix} 1 & 0 \\ 0 & 1 \end{pmatrix} Z + \begin{pmatrix} -H_2^+ & -H_3^+ W \\ H_3^+ W & 0 \end{pmatrix} + (H_4^+ x_2 + H_5^+ y_2) \begin{pmatrix} \dfrac{\partial Z}{\partial x_2} & -\dfrac{\partial Z}{\partial x_1} \\ -\dfrac{\partial Z}{\partial x_1} & -\dfrac{\partial Z}{\partial x_2} \end{pmatrix}. \tag{2}$$

Für $x_2 > 0$, $y_2 < 0$ ist in (2) H_k^+ durch H_k^- ($k = 1, 2, 4, 5$) und H_3^+ durch $-H_3^-$ zu ersetzen.

Die Konstanten $H_1^+, \dots, H_{11}^+, H_1^-, \dots, H_{11}^-$ sind dieselben wie in [2]. Sie seien der Vollständigkeit halber hier noch einmal mitgeteilt:

$$H_1^- = \frac{\mu_1(\lambda_0 + 3\mu_0)(\lambda_1 + 2\mu_1) + \mu_0(\lambda_1 + 3\mu_1)(\lambda_0 + 2\mu_0)}{[\mu_0(\lambda_1 + 3\mu_1) + \mu_1(\lambda_1 + \mu_1)][\mu_1(\lambda_0 + 3\mu_0) + \mu_0(\lambda_0 + \mu_0)]},$$

$$H_2^- = H_1^- - \frac{2}{\mu_0 + \mu_1},$$

$$H_3^- = \frac{\mu_0^2(\lambda_1 + 3\mu_1) - \mu_1^2(\lambda_0 + 3\mu_0)}{[\mu_1(\lambda_0 + 3\mu_0) + \mu_0(\lambda_0 + \mu_0)][\mu_0(\lambda_1 + 3\mu_1) + \mu_1(\lambda_1 + \mu_1)]},$$

$$H_4^- = -\frac{\lambda_1 + \mu_1}{\mu_0(\lambda_1 + 3\mu_1) + \mu_1(\lambda_1 + \mu_1)},$$

$$H_5^- = \frac{\lambda_0 + \mu_0}{\mu_1(\lambda_0 + 3\mu_0) + \mu_0(\lambda_0 + \mu_0)},$$

$$H_6^- = H_1^- - \frac{\lambda_0 + 3\mu_0}{2\mu_0(\lambda_0 + 2\mu_0)},$$

$$H_7^- = H_6^- + \frac{\mu_1 - \mu_0}{\mu_0(\mu_0 + \mu_1)},$$

$$H_8^- = H_3^-,$$

$$H_9^- = \frac{(\mu_0 - \mu_1)(\lambda_0 + \mu_0)(\lambda_0 + 3\mu_0)}{2\mu_0(\lambda_0 + 2\mu_0)[\mu_1(\lambda_0 + 3\mu_0) + \mu_0(\lambda_0 + \mu_0)]},$$

$$H_{10}^- = H_9^-,$$

$$H_{11}^- = \frac{(\lambda_0 + \mu_0)^2 (\mu_1 - \mu_0)}{\mu_0(\lambda_0 + 2\mu_0)[\mu_1(\lambda_0 + 3\mu_0) + \mu_0(\lambda_0 + \mu_0)]}.$$

H_k^+ erhält man aus H_k^-, indem man bei den Laméschen Moduln λ_i, μ_i die Indizes 0 und 1 vertauscht.

$G(x, y)$ hat für $y_2 \neq 0$ folgende Eigenschaften, durch die $G(x, y)$ vollständig bestimmt ist:

$$\underset{(1)}{A}(\partial_x)\, V(x, y) = 0 \quad \text{für } x_2 > 0, \qquad \underset{(0)}{A}(\partial_x)\, V(x, y) = 0$$

für $x_2 < 0$,

$$\lim_{x_2 \to +0} G(x, y) = \lim_{x_2 \to -0} G(x, y),$$

$$\lim_{x_2 \to +0} \underset{(1)}{T}(\partial_x, (0, 0, 1))\, G(x, y) = \lim_{x_2 \to -0} \underset{(0)}{T}(\partial_x, (0, 0, 1))\, G(x, y).$$

Hierbei ist $\underset{(i)}{T}(\partial_x, n(x))$ der zweidimensionale Spannungsoperator mit den Komponenten

$$\underset{(i)}{T_{jk}}(\partial_x, n(x)) = \lambda_i n_j(x) \frac{\partial}{\partial x_k} + \mu_i n_k(x) \frac{\partial}{\partial x_j} + \mu_i \delta_{jk} \sum_{l=1}^{2} n_l(x) \frac{\partial}{\partial x_l}.$$

Es ist $G(x, y)$ symmetrisch.

Bezeichnet $E_0 = \dfrac{\mu_0(2\mu_0 + 3\lambda_0)}{\lambda_0 + \mu_0}$ den linearen Elastizitätsmodul der unteren Halbebene, dann erhält man durch die Grenzübergänge

$$\lim_{E_0 \to \infty} G(x, y) = G_I(x, y) \quad \text{für } x_2 > 0, \; y_2 > 0,$$

$$\lim_{E_0 \to 0} G(x, y) = G_{II}(x, y) \quad \text{für } x_2 > 0, \; y_2 > 0$$

die bekannten elastostatischen Greenschen Tensoren 1. Art $G_I(x, y)$ (Verschiebung am Rand Null) und 2. Art $G_{II}(x, y)$ (Normalspannung am Rand Null) für die Halbebene (s. etwa [4], S. 41–43).

Analog wie in [1], S. 119 kann man das Problem thermischer Einschlüsse behandeln (Problem W). Befinden sich in den Halbebenen $H_1 = \{x: x_2 > 0\}$, $H_0 = \{x: x_2 < 0\}$ Gebiete $D_i^j \subset H_i$, $i = 0, 1$, $j = 1, \dots, p_i$, mit konstanter Temperatur Θ_i^j, dann gilt für den Verschiebungsvektor u

$$2\pi u(x) = \sum_{i=0,1} \sum_{j=1,\dots,p_i} (2\mu_i + 3\lambda_i)\,\alpha_i \Theta_i^j \int_{D_i^j} \mathrm{Div}_y G(x, y)\, \mathrm{d}y\,. \quad (3)$$

Dabei ist

$$\mathrm{Div}_y\, G(x,y) = \begin{pmatrix} \dfrac{\partial}{\partial y_1} G_{11}(x, y) + \dfrac{\partial}{\partial y_2} G_{12}(x, y) \\ \dfrac{\partial}{\partial y_1} G_{21}(x, y) + \dfrac{\partial}{\partial y_2} G_{22}(x, y) \end{pmatrix}.$$

Nun ist für $x_2 > 0$, $y_2 > 0$

$$\mathrm{Div}_y\, G(x, y) = \frac{1}{\lambda_1 + 2\mu_1} \begin{pmatrix} \dfrac{\partial}{\partial x_1} \ln |x - y| \\ \dfrac{\partial}{\partial x_2} \ln |x - y| \end{pmatrix} + $$

$$+ (H_6^+ + H_8^+ - H_{10}^+) \begin{pmatrix} \dfrac{\partial Z}{\partial x_1} \\ -\dfrac{\partial Z}{\partial x_2} \end{pmatrix} + (2H_9^+ + H_{11}^+)\, x_2 \begin{pmatrix} \dfrac{\partial^2 Z}{\partial x_1\, \partial x_2} \\ \dfrac{\partial^2 Z}{\partial x_2^2} \end{pmatrix}$$

(für $x_2 < 0$, $y_2 < 0$ steht λ_0, μ_0, H_6^-, H_8^-, H_9^-, H_{10}^-, H_{11}^-) und für $x_2 < 0$, $y_2 > 0$

$$\mathrm{Div}_y\, G(x, y) = (H_1^+ + H_3^+ - H_5^+) \begin{pmatrix} \dfrac{\partial Z}{\partial x_1} \\ \dfrac{\partial Z}{\partial x_2} \end{pmatrix}$$

(für $x_2 > 0$, $y_2 < 0$ steht H_1^-, H_3^-, H_5^-).
Das Problem W ist also explizit lösbar, wenn man das logarithmische Potential der Gebiete D_i^j kennt. Die Formel (3) gibt auch die Kontaktbedingungen richtig wieder, wenn der Durchschnitt des Randes von $\cup D_i^j$ mit der Geraden $x_2 = 0$ ein positives Maß hat (s. [5]).

Literatur

1 JENTSCH, L., Über Wärmespannungen in Körpern mit stückweise konstanten LAMÉschen Elastizitätsmoduln, Schriftenreihe des Zentralinstituts für Math. und Mechanik bei der AdW der DDR, Heft 14, 1972, Akademie-Verlag Berlin.

2 JENTSCH, L., Zur Theorie der Wärmespannungen in Bimetallkörpern, Wiss. Z. Karl-Marx-Univ. Leipzig, Math.-Naturw. R., 29. Jg. (1980), H. 1, S. 49–58.

3 MAUL, J., Eine einheitliche Methode zur Lösung der ebenen Aufgaben der linearen Elastostatik, Schriftenreihe des Zentralinstituts für Math. und Mechanik bei der AdW der DDR, Heft 24, 1976, Akademie-Verlag Berlin.

4 Ворович, И. И., Александров, В. М. Ъабешко, В. А. Неклассические смешанные задачи теории упругости, Издательство Наука, Москва 1974

5 HEYNIG, I., Über Wärmespannungen in gekoppelten elastischen Halbräumen mit speziellen unstetigen Temperaturfeldern, Diplomarbeit, Sektion Mathematik Technische Hochschule Karl-Marx-Stadt, 1979.

Eingereicht am 4. 11. 1980

Anschrift: Prof. Dr. LOTHAR JENTSCH, TH Karl-Marx-Stadt, DDR-9010 Karl-Marx-Stadt, Postschließfach 964

BUCHBESPRECHUNGEN

Levin, M. / Girshovich, J., Optimal Quadrature Formulas. Teubner-Texte zur Mathematik. Leipzig, BSB B.G. Teubner Verlagsgesellschaft 1979. 122 S., M 13,–. BN 6659572

Eine Verallgemeinerung des arithmetisch-geometrischen Mittels

Für das Problem, geeignete Koeffizienten A_{kj}, x_k für eine Quadraturformel

$$\int_0^1 f(x)\, \mathrm{d}x = \sum_{k=1}^{n} \sum_{j \in J_k} A_{kj} f^{(j)}(x_k) + R(f)$$

zu finden, gibt es zwei verschiedene Ansätze: Erstens kann man $f(x)$ durch Polynome approximieren (TAYLOR, WEIERSTRASS) und $R(f) = 0$ für Polynome möglichst hohen Grades fordern (z. B. Gaußsche Formeln), zweitens kann man eine Funktionenmenge H vorgeben und die Koeffizienten so bestimmen, daß

$$R(H) = \sup_{f \in H} |R(f)|$$

möglichst klein wird. Diesem zweiten Ansatz ist das vorliegende Buch in großer Ausführlichkeit gewidmet. Die betrachteten Funktionenmengen H sind meist vom Typ

$$W^r L_q = \{f \mid f \text{ ist } r\text{-mal differenzierbar, } \int_0^1 |f^{(r)}(x)|^q\, \mathrm{d}x \leqq 1\}$$

mit eventuellen zusätzlichen Randbedingungen, Periodizitätseigenschaften usw.
Formeln, für welche bei vorgegebenen Knoten $x_1, \dots, x_n$ die Gewichte A_{kj} so bestimmt werden, daß $R(H)$ minimal wird, werden *beste* Formeln genannt (SARD), Formeln bei welchen sowohl über die x_k als auch die A_{kj} optimiert wird, heißen optimal (NIKOLSKY, SCHOENBERG). Für eine dritte Problemklasse wird $x_1 = 0$ und/oder $x_n = 1$ verlangt und die restlichen x_k bleiben frei (MARKOV 1885, RADAU 1880). Über die PEANOsche Fehlerformel und partielle Integration hängt die Konstruktion „bester" Formeln mit der Theorie der Splinefunktionen zusammen (SCHOENBERG), was in großer Allgemeinheit im fundamentalen „Theorem 1.6.1" die Basis des Buches bildet. Anschließend werden dann die Berechnung optimaler Formeln für eindimensionale Integration, und Formelklassen, welche für $n \to \infty$ optimal sind, ausführlich besprochen.

Auf die praktische Nützlichkeit der Theorie, insbesondere den Vergleich mit Algorithmen welche die Stützstellen in Abhängigkeit der zu integrierenden Funktion f anpassen (etwa von BACHVALOV), wird nicht eingegangen. Ein ausführliches Literaturverzeichnis beschließt das Buch.

Genève G. WANNER

Maess, G., Iterative Lösung linearer Gleichungssysteme. Nova Acta Leopoldina NF 238/Bd. 52. Halle, Deutsche Akademie der Naturforscher Leopoldina 1979. 79 S., M 24,–.

Nach einer knappen Übersicht über die iterativen Methoden zur Lösung von N linearen Gleichungen in N Unbekannten und deren Konvergenzverhalten wendet sich der Verfasser dem Problem der iterativen Lösung von rechteckigen Systemen zu, die in der Rechenpraxis eine nicht zu unterschätzende Rolle spielen. Zunächst werden die stationären und instationären Verfahren ganz allgemein betrachtet und die notwendigen Konvergenzaussagen zusammengestellt. Konkret werden dann die Spaltenapproximation, d. h. die Konstruktion einer besten Approximation durch sukzessive Bestimmung von besten Approximationen in Teilräumen, und die Methode der Projektion auf

Schnitträume von Hyperebenen behandelt und deren Eigenschaften eingehend dargestellt. Sehr ausführliche Literaturhinweise ergänzen das interessante Buch, in welchem wohl zum ersten Mal der Stand der Rechentechnik auf diesem Spezialgebiet in Buchform enthalten ist.

Zürich H-R. SCHWARZ

Codes for Boundary-Value Problems in Ordinary Differential Equations. Proceedings of a Working Conference, May, 1978. Edited by Childs, B. / Scott, M. / Daniel, J. W. / Denman, E. / Nelson, P., Lecture Notes in Computer Science 76. Berlin-Heidelberg-New York, Springer-Verlag 1979. VIII, 388 S., DM 35,50. US $ 19.60. ISBN 3-540-09554-3

Die numerische Lösung von Randwertaufgaben mit gewöhnlichen Differentialgleichungen ist eines der Hauptanliegen der numerischen Mathematik und wird gegenwärtig in vielfältiger Weise untersucht. Große Fortschritte bei der Behandlung der Randwertprobleme wurden erzielt, nachdem es gelungen war, die Lösung von Anfangswertaufgaben auf einen hohen Stand zu bringen. Bei der Lösung der Randwertprobleme haben sich im wesentlichen die folgenden Verfahren bewährt. Die Randwertaufgaben werden mittels mehrfacher Schießmethoden, Arbeiten u. a. von H. B. KELLER, M. R. OSBORNE, R. BULIRSCH, J. STOER, P. DEUFLHARD, auf Anfangswertaufgaben zurückgeführt, wobei zur Erfüllung der Schießbedingungen im allgemeinen nichtlineare Gleichungssysteme zu lösen sind. Bei einem weiteren bewährten Verfahren werden die Differentialgleichungen mittels der Trapezregel diskretisiert und die entstehenden nichtlinearen Gleichungen gelöst. Wesentliche Beiträge dazu sind u. a. in den Arbeiten von V. PEREYRA, M. LENTINI und H. B. KELLER zu finden. Andere Lösungsverfahren benutzen die Möglichkeiten der Superposition, Orthonormalisierung und der Quasilinearisierung der Randwertaufgaben oder die Methode der Kollokation. Auf der Arbeitskonferenz standen vorwiegend die für die Lösung der Randwertaufgaben entwickelten Programme und verschiedene Programmpakete hierzu im Vordergrund, einschließlich der Darlegung der numerischen Erfahrungen und der Leistungsfähigkeit der Programme. Der Konferenzbericht enthält eingeladene Arbeiten, ihre Autoren sind J. W. DANIEL, H. A. WATTS, P. DEUFLHARD, V. PEREYRA, M. R. SCOTT, F. GLADWELL, U. ASCHER, J. CHRISTIANSEN, R. D. RUSSEL, B. CHILDS, H. R. PORTER, G. W. REDDIEN und E. D. DENMAN, mitgeteilte Arbeiten und die Zusammenfassung von mehreren Workshops.

Der Bericht gibt einen sehr guten Überblick über den gegenwärtigen Stand der numerischen Lösung von Randwertaufgaben mit gewöhnlichen Differentialgleichungen und der dazu zu verwendenden Verfahren, Programme und Programmpakete.

Berlin F. GRUND

Parlett, B. N., The Symmetric Eigenvalue Problem. Prentice-Hall Series in Computational Mathematics. Englewood Cliffs, N.J. 07632, Prentice-Hall, Inc. 1980. XX, 348 S., $ 25.00. ISBN 0-13-880047-2

Mit dem Titel „The Symmetric eigenvalue problem" legt der international bekannte Numeriker B. N. PARLETT ein Buch mit teilweise Lehrbuch- und Monographiecharakter vor, das der numerisch arbeitende Mathematiker schnell als interessante und aktuelle Weiterführung des Standartwerkes „The algebraic eigenvolue problem" von J. H. WILKINSON aus dem Jahre 1965 schätzen lernt. Von der Thematik her ist es gegenüber dem WILKINSONschen Standartwerk eingeschränkt und nur auf das symmetrische Eigenwertproblem konzentriert, doch es erfüllt im wesentlichen die vom Autor selbst formulierte Hoffnung, daß jedes Kapitel gerade auch für den Spezialisten etwas Interessantes enthalten möge. Das Buch umfaßt 15 Kapitel, wobei im ersten Kapitel die Grundbegriffe symmetrischer Matrizen und Geometrie endlichdimensionaler linearer Räume mit innerem Produkt bzw. Norm kurz und übersichtlich zusammengestellt sind.

Im Kapitel 2 stellt der Autor die jedem Numeriker elektrisierende Frage „Was ist klein? Was ist groß?" Außerdem beschreibt er hier die im Buch benutzte Methode der Rundungsfehlerabschätzung. Im Kapitel 3 wird Matrixfaktorisierung in obere und untere Dreiecksmatrix unter Beachtung der Spektrumsverschiebung betrachtet. Das Kapitel 4 ist der Vektoriteration gewidmet. In diesem Kapitel wird insbesondere auch die RAYLEIGH-Quotient-Iteration beschrieben und ihre lokale kubische Konvergenz bewiesen. Neben diesem wohlbekannten Ergebnis werden weitere interessante Fakten hergeleitet, die neben einer tiefgreifenden theoretischen Fundierung auch die effektive Nutzung begründen. Das nur wenige Seiten umfassende Kapitel 5 enthält die Beschreibung des Prinzips der Deflation einer Matrix, d. h. der Ordnungsreduzierung, wenn einzelne Eigenwerte und Eigenvektoren bekannt sind. Kapitel 6 ist den orthogonalen Matrizen (JACOBI- und GIVENSrotationen) und den damit verbundenen Faktorisierungsmethoden (QR-Faktorisierung, GRAM-SCHMIDT-Verfahren) gewidmet. In den Kapitel 7 bis 9 erfolgt eine eingehende Beschreibung der Herstellung von tridiagonalen Matrixformen, der QL- und QR-Algorithmen und des JACOBI-Verfahrens. Das methodische Vorgehen in diesen Kapiteln wird durch Beschreibung des Verfahrens, allgemeines Konvergenzverhalten, Betrachtung von Sonderfällen und Fehleruntersuchungen charakterisiert.

Die ersten neun Kapitel des Buches bilden eine relative Einheit und umfassen Algorithmen zur iterativen Berechnung von Eigenwerten und Eigenvektoren von Matrizen. Das zehnte Kapitel leitet zu spezifischen Fragen über, die nicht immer als direkte Fortsetzung oder Verallgemeinerung des Inhalts der ersten Kapitel angesehen werden können. Im Kapitel 10 werden Eigenwertschranken angegeben, die auf der Basis von Informationen über Untermatrizen gewonnen werden. Neben der klassischen min-max-Carakterisierung nehmen die optimalen Intervalle von N. J. LEHMANN einen breiten Rahmen ein. Von einer gewissen Originalität ist Kapitel 11, in dem das RITZ-Verfahren zur Eigenwertberechnung auf den Fall von Matrizen angewandt wird; beschrieben werden die entsprechenden Unterraumtechniken. In Kapitel 12 wird der Begriff des KRYLOW-Unterraums als die Aufspannung der Vektoren f, Af, ..., $A^{m-1}f$ mit einer quadratischen Matrix A definiert. Dieser Unterraum wird in Zusammenhang mit Kapitel 11 untersucht. Schließlich wird im Kapitel 13 gezeigt, daß die Anwendung der RITZ-Technik auf den KRYLOW-Unterraum das bekannte LANCZOS-Verfahren ergibt. Wird in der KRYLOW-Folge f, Af, ..., $A^{m-1}f$ der Vektor f durch einen Unterraum ersetzt, so ergibt sich eine Folge von Unterräumen. In Kapitel 14 wird die Darstellung dieser Folge und ihre Nützlichkeit untersucht. Schließlich wird in Kapitel 15 das allgemeine Eigenwertsproblem behandelt. Viele der in den vorangegangenen Kapiteln behandelten Algorithmen und Aussagen werden für diese Situation angepaßt bzw. modifiziert. In zwei kurzen Anhängen werden noch Rang-1-Matrizen und CHEBYSHEW-Polynome dargelegt.

Die Lesbarkeit und Verständlichkeit der einzelnen Kapitel wird durch illustrative Beispiele erhöht, außerdem beeindruckt die straffe Darlegung des gesamten Material und die kompakte Beweisführung. Der notwendige erläuternde Kommentar ist knapp auf das wesentliche beschränkt.

Karl-Marx-Stadt F. KUHNERT

Lehrbuch Höhere Festigkeitslehre. Band 1: Grundlagen der Höheren Festigkeitslehre. Autorenkollektiv unter Federführung von Göldner, H., Leipzig, VEB Fachbuchverlag 1979. 236 S., 116 Abb., M 16,—. Auslandspreis 24,—. BN 5464343

Der vorliegende erste Band des Lehrbuches „Höhere Festigkeitslehre" ist vorzugsweise als Studienhilfe für Studierende einiger Fachrichtungen der Grundstudienrichtung Maschinenwesen gedacht, die aufbauend auf dem Grundkurs in Technischer Mechanik eine weiterführende Ausbildung in Festigkeitslehre erfahren.

Im einzelnen werden folgende Kapitel behandelt: Grundlagen der Elastizitätstheorie bis zu den elastischen Grundgleichungen; Ebener Spannungs- u. Verzerrungszustand mit der Darstellung in Karthesischen und Polarkoordinaten und zahlreichen klassischen Lösungen wie die Halbebene unter Einzellast, Rechteckscheibe unter äußerer Belastung, rotierende Scheiben, Kreisbogenträger, HERTZsche Pressung usw.; KIRCHHOFFsche Plattentheorie mit Anwendungen auf Beispiele in Karthesischen und Polarkoodinaten für nicht rotationssymmetrische Belastung.

In diesen ersten drei Abschnitten wird das Verhalten unter Temperaturbelastung mit betrachtet.

Membrantheorie rotationssymmetrisch belasteter Drehschalen und die Biegetheorie der Kreiszylinderschale unter rotations-symmetrischer Belastung; Freie Torsion prismatischer Stäbe mit Voll- und dünnwandigen mehrfach zusammenhängenden Querschnitten sowie die Wölbkrafttorsion offener dünnwandiger Profile; Querkraftschubspannungen offener u. mehrfach zusammenhängender Querschnitte; Drehsymmetrische Beanspruchungen von Rotationskörpern (Halbraum unter Einzellast (Boussinesq), Torsion nichtprismatischer Stäbe, dicke Platten); Statik der Stabtragwerke mit den insbesondere im Bauwesen und Stahlbau üblichen Methoden (Kraftgrößen- u. Verrückungsmethode).

Jeder Abschnitt enthält am Ende einige Fragen zur Selbstkontrolle. Es wird sich mit Absicht auf die „klassischen" Methoden der Elastizitätstheorie beschränkt, die eine geschlossene analytische, anschauliche Darstellung des Verhaltens einfacher elastischer Systeme erlauben. Die numerische Behandlung der Aufgaben bleibt anderen Büchern vorbehalten. Trotz des fünfköpfigen Autorenkollektivs ist eine einheitliche Darstellung bei wenig unterschiedlichem Stil gelungen.

Es wird sowohl als Lehrbuch dem Studierenden, der anhand des 1978 im gleichen Verlag erschienen „Arbeitsbuch Höherer Festigkeitslehre" den Stoff durch Übung vertiefen kann, als auch dem praktisch tätigen Berechnungsingenieur und Konstrukteur eine wertvolle Hilfe sein.

Bei einer Neuauflage sollten die Bilder 3.3, 4.2, 4.7 und 7.3 ergänzt bzw. durch bessere ersetzt werden.

Merseburg W. Pfefferkorn

Hudson, J. A., The Excitation and Propagation of Elastic Waves. Cambridge Monographs on Mechanics and Applied Mathematics. Cambridge, Cambridge University Press 1980. VIII, 226 S., £ 14.00 A H/C. ISBN 0-521-22 777-1

Das Buch behandelt die Grundlagen der linearisierten Elastodynamik vom Standpunkt der Wellenausbreitung in einem isotropen, unbegrenzten Körper. Hinsichtlich der Anwendung ist an solche Gebiete wie Seismologie, Lärmanalyse oder die zerstörungsfreie Werkstoffprüfung gedacht. Das einheitliche Kapitel über die linearisierten Bewegungsgleichungen setzt Grundkenntnisse in der Elastizitätstheorie voraus und stellt solche Aspekte in den Vordergrund, die für die dynamische Theorie von Wichtigkeit sind. Im zweiten Kapitel werden ebene, sphärische und zylindrische Wellen als fundamentale Lösungen der elastodynamischen Gleichungen beschrieben, die für ein homogenes, unendliches Material als Kompressions- oder Schubwellen mit ihren unterschiedlichen Wellengeschwindigkeiten charakteristisch sind. Es folgen die Oberflächen- und Grenzflächenwellen (Stonley-, Love- und Rayleighwellen), die Reflexion und Brechung ebener Wellen und das elastodynamische Äquivalent des Snelliusschen Brechungsgesetzes. Der zum klassischen Bestand der Elastodynamik zu rechnende Stoff wird durch Ergebnisse der letzten Jahrzehnte ergänzt. Dazu gehört der theoretische Nachweis, daß auch bei inhomogenem Material zwei Wellenfronten mit unterschiedlichen Geschwindigkeiten existieren. In einem Kapitel über das allgemeine Anfangs- und Randwertproblem werden die Grundlagen der Beugungstheorie beschrieben, die auf dem elastodynamischen Äquivalent des Greenschen Theorems beruhen. Dazu werden Fragen der Eindeutigkeit und Reziprozität behandelt. Das Buch setzt fort mit einer Beschreibung harmonischer Probleme, bei denen also die Zeitabhängigkeit aller Störungsquellen harmonisch ist und sich nach einer gewissen Zeit stationäre Zustände eingestellt haben und behandelt die zum Beweis der Eindeutigkeit erforderlichen Bedingungen im Unendlichen (nach Kupradze 1963). Die Fragen der Wechselwirkung elastischer Wellen mit Grenzflächen stehen in dem Buche nicht im Vordergrund. Sie werden jedoch in einem Kapitel über eine Linienquelle im homogenen Halbraum angeschnitten. Dabei wird insbesondere beschrieben, wie durch die Wechselwirkung von Zylinderwellen mit einer ebenen, freien Oberfläche Oberflächenwellen erzeugt werden. Die Greensche Funktion für den Halbraum wird aufgestellt und das Lambsche Problem betrachtet. Das letzte Kapitel ist der Wellenausbreitung in linear viskoelastischen Werkstoffen gewidmet, bei denen die Spannungen von der Dehnungsgeschichte abhängen und die Wellen eine Dämpfung aufweisen. Insgesamt vermittelt das Buch einen guten Überblick über die Ergebnisse und Methoden der linearen Elastodynamik.

Woltersdorf G. Haberland

In der DDR beziehbar im Rahmen des Kontingents.

EINGEGANGENE BÜCHER

Bei der Schriftleitung sind folgende Bücher eingegangen (ausführliche Besprechung bleibt vorbehalten).

Jaffe, A. / Taubes, C., Vortices and Monopoles, Structure of Static Gauge Theories, Progress in Physics 2, Boston-Basel-Stuttgart, Birkhäuser Verlag 1980, 287 S., DM 30,–. ISBN 3-7643-3025-2

Meschkowski, H., Mehrsprachenwörterbuch mathematischer Begriffe, deutsch-englisch-französisch-russisch-italienisch, Mannheim-Wien-Zürich, B. I.-Wissenschaftsverlag 1972, 135 S., ISBN 3-411-01409-1

Lüst, R., Hydrodynamic, Mannheim-Wien-Zürich, B. I.-Wissenschaftsverlag 1978, 234 S., ISBN 3-411-01540-3

de Groot, S. R., Thermodynamik irreversibler Prozesse, HTB 18/18a, Mannheim, B. I.-HTB-Verlag 1960, XV, 216 S.

Mahrenholtz, O., Analogrechnen in Maschinenbau und Mechanik, HTB 154/154a, Mannheim-Zürich, B. I.-HTB-Verlag 1968, 208 S.

de Groot, S. R. / Mazur, P., Anwendung der Thermodynamik irreversibler Prozesse, Mannheim-Wien-Zürich, B. I.-Wissenschaftsverlag 1974, 349., ISBN 3-411-01449-0

Marguerre, K. / Wölfel, H., Technische Schwingungslehre, Lineare Schwingungen vielgliedriger Strukturen, Mannheim-Wien-Zürich, B. I.-Wissenschaftsverlag 1979, 338. S., ISBN 3-411-01562-4

Sandford, D. M., Using Sophisticated Models in Resolution Theorem Proving, Lecture Notes in Computer Science 90, Berlin-Heidelberg-New York, Springer-Verlag 1980, 239 S., 6 Abb., DM 25,–. US $ 14.70. ISBN 3-540-10231-0

Wood, D., Grammer and L Forms: An Introduction, Lecture Notes in Computer Science 91, Berlin-Heidelberg-New York, Springer-Verlag 1980, IX, 314 S., DM 32,80, US $ 19.30. ISBN 3-540-10233-7

Hafer, X. / Sachs, G., Flugmechanik, Moderne Flugzeugentwurfs- und Steuerungskonzepte, Hochschultext, Berlin-Heidelberg-New York, Springer-Verlag 1980, XVIII, 263 S., 155 Abb., DM 58,–. US $ 34.10. ISBN 3-540-10072-5

Eppler, R. / Fasel, H. (Eds.), Laminar-Turbulent Transition, IUTAM Symposium Stuttgart/Germany 1979, Berlin-Heidelberg-New York, Springer-Verlag 1980, XVIII, 432 S., 289 Abb., DM 74,–, US $ 43.60. ISBN 3-540-10142-X

Kall, P. / Prékopa, A. (Eds.), Recent Results in Stochastic Programming, Proceedings, Oberwolfach 1979, Lecture Notes In Economics and Mathematical Systems 179, Berlin-Heidelberg-New York, Springer-Verlag 1980, IX, 237 S., 10 Abb., DM 34,–, US $ 20.10. ISBN 3-540-10013-X

Dembiński, P. (Ed.), Mathematical Foundations of Computer Science 1980, Proceedings of the 9th Symposium, Rydzyna, Poland, 1980. Lecture Notes in Computer Science 88, Berlin-Heidelberg-New York, Springer-Verlag 1980, VIII, 723 S., 99 Abb., DM 63,–, US $ 37.20. ISBN 3-540-10027-X

Kavanau, J. L., Symmetry: An Analytical Treatment, Los Angeles/California, Science Software Systems, Inc. 1980. ca. 200 S., ca. 500 Abb., $ 10.00. ISBN 0-937292-00-1

Collet, P. / Eckmann, J.-P., Iterated Maps on the Interval as Dynamical Systems, Progress in Physics 1, Basel-Boston-Stuttgart, Birkhäuser Verlag 1980, VII, 248 S., DM 30,–, ISBN 3-7643-3026-0

NACHRICHTEN

WACŁAW OLSZAK † 8. 12. 1980

Am 8. Dezember 1980 verstarb in Udine (Italien), wo er in den letzten elf Jahren Rektor des Internationalen Zentrums der Mechanischen Wissenschaften (CISM) war, der hervorragende polnische Gelehrte WACŁAW OLSZAK. Geboren wurde er am 24. Oktober 1902 in Karvina im ehemaligen Herzogtum Teschen, wo seine Familie eng mit der polnischen Nationalbewegung verbunden war. Von 1920 bis 1925 studierte er an der Technischen Hochschule Wien Tief- und Wasserbau, wobei er gleichzeitig starke Interessen am kulturellen Leben entwickelte und u. a. am Wiener Konservatorium die Klasse Violine absolvierte. In den Jahren 1926 und 1927 setzte er seine Studien an der Faculté des Sciences der Pariser Universität fort und widmete sich dabei im besonderen Fragen der ingenieurmäßigen Anwendung der Mathematik und der theoretischen und angewandten Mechanik. Nach seiner Rückkehr nach Polen arbeitete er am Bau von Brücken und Staudämmen, im Bergbau und in der Hüttenindustrie. Schon damals veröffentlichte er mehrere Arbeiten, die erste erschien sogar schon 1922.

Eine langwierige Krankheit zwang WACŁAW OLSZAK zu einem jahrelangen Sanatoriumsaufenthalt. Er arbeitete damals an zwei Dissertationen. Die erste — zur Theorie des Stahlbetons und der Stahlbaukonstruktionen — verteidigte er 1933 an der Technischen Hochschule Wien, die zweite — unter Anleitung von M. T. HUBER über Fragen der mathematischen Elastizitätstheorie geschriebene — 1934 an der Technischen Hochschule Warschau. Im Jahre 1937 habilitierte sich WACŁAW OLSZAK an der Berg- und Hüttenakademie in Kraków mit einer Arbeit über die Statik und Dynamik von Luftschutzkonstruktionen und wurde dort Privatdozent. Im zweiten Weltkrieg wurde seine Familie verfolgt; er verlor seine Eltern und die materielle Existenzgrundlage. Er war als Arbeiter und Kraftfahrer tätig, verbrachte nach Kriegsende einige Monate in Frankreich und fast ein ganzes Jahr in Belgien und trug in dieser Zeit an der Faculté des Sciences Appliquées in Liège vor. Berufungen auf Lehrstühle in München, Wien und Dresden lehnte er ab.

Im Jahre 1946 kehrte WACŁAW OLSZAK nach Polen zurück und übernahm als Ordinarius den Lehrstuhl für Werkstoffestigkeit an der Berg- und Hüttenakademie Kraków und leitete hier und an der neu gegründeten Technischen Hochschule in Kraków Untersuchungen über vorgespannte Konstruktionen und andere Fragen der angewandten Mechanik.

Seit 1952 war WACŁAW OLSZAK eng mit der Technischen Hochschule Warschau und mit der in dieser Zeit gegründeten Polnischen Akademie der Wissenschaften (PAN) verbunden. Er gründete den ersten Lehrstuhl für Elastizitäts- und Plastizitätstheorie in Polen und organisierte ein wissenschaftliches Forschungszentrum für Festkörpermechanik. Von 1963 bis 1970 war er Direktor des Instituts für Grundlagen der Technik der PAN und daneben viele Jahre Mitglied des Präsidiums und Sekretär der Abteilung für Technische Wissenschaften der PAN. Sein Hauptarbeitsgebiet war von Anfang an die Elastizitäts- und Plastizitätstheorie. Zu seinen grundlegenden Untersuchungen gehört die Entwicklung der Theorie anisotroper wie auch der inhomogener Medien. Anwendungen seiner Forschungsergebnisse betreffen die Mechanik plastischer Konstruktionen. Seine Initiative und Mitwirkung waren von grundlegender Bedeutung bei der Untersuchung verschiedener Problemkreise der Platten- und Schalentheorie in Polen. In den Rahmen der Plastizitätstheorie gehören auch die von ihm gefundenen Lösungen von Problemen der Mechanik von Granulaten mit den bei diesen vorhandenen Inhomogenitäten. Der Theorie anisotroper und inhomogener Medien widmete er gemeinsam mit seinen Mitarbeitern eine Reihe von Monographien. Ein weiteres Wissensgebiet, das seine Entwicklung in Polen der Forschungsarbeit und den Monographien von WACŁAW OLSZAK verdankt, ist die Theorie der vorgespannten Konstruktionen und damit verbunden die Rheologie.

Mehrere Arbeiten widmete WACŁAW OLSZAK den Fragen der Popularisierung der Wissenschaft, dem Zusammenwirken der einzelnen Zweige der Mechanik, der historischen Entwicklung der Mechanik und der Perspektive ihrer Teilgebiete.

Die vielseitige schöpferische Tätigkeit WACŁAW OLSZAKS, in deren Rahmen er auch eine Gruppe von Schülern heranbildete, die seine Konzeptionen weiterentwickeln und die Mechanik durch neue Grundlagen- und angewandte Forschungen bereichern, fanden nicht nur in Polen Anerkennung. Seine Bücher wurden ins Englische, Russische, Tschechische und Rumänische übersetzt. Zahlreiche Hochschulen des In- und Auslandes wie die Universitäten bzw. Technischen Hochschulen in Wien, Dresden, Toulouse, Liège, Glasgow, Warschau und Kraków (hier auch die Berg- und Hüttenakademie) verliehen ihm die Ehrendoktorwürde. Er war Mitglied vieler ausländischen Akademien, u. a. der Französischen Akademie der Wissenschaften, der Königlichen Akademie der Technischen Wissenschaften in Stockholm, der Akademie der Technischen Wissenschaften in Helsinki, der Ungarischen, der Bulgarischen, der Argentinischen, der Luxemburgischen, der Slovenischen, der Österreichischen Akademie der Wissenschaften und der Deutschen Akademie der Naturforscher Leopoldina zu Halle/S. Von den zahlreichen Auszeichnungen, die er erhielt, seien erwähnt: die Ehrenmedaillen der Universiäten Liège (1946) und Neapel (1967) und der Československá Akademie Věd, Prag (1967), das Österreichische Ehrenzeichen für Wissenschaft und Kunst (1964), das Offizierskreuz des Französischen Verdienstordens für Wissenschaft und Erfindung (1955) und zweimal der Nationalpreis I. Klasse der Volksrepublik Polen (1955 und 1966).

Im internationalen wissenschaftlichen Leben war WACŁAW OLSZAK außerordentlich aktiv. Er war u. a. Mitbegründer der Réunion Internationale des Laboratoires d'Essais et de Recherches sur les Matériaux et les Constructions (RILEM) in Paris (1947) und von 1962 bis 1964 deren Präsident, Mitbegründer der International Association for Shell Structures (IASS) in Oslo (1957) und deren Vizepräsident, Vertreter Polens in der International Union for Theoretical and Applied Mechanics (IUTAM) und über 20 Jahre Mitglied des Büros dieser Organisation, Mitglied des Ständigen Ausschusses der Internationalen Vereinigung für Brückenbau und Hochbau (IVBH) und Mitglied der Gesellschaft für Angewandte Mathematik und Mechanik (GAMM). Als Experte der UNESCO übte er in Nepal eine Beratertätigkeit bei der Gründung und der Einrichtung von „Engineering Colleges" an der Tribhuwan University in Katmandu aus (1966). An der Organisation und Einrichtung des Internationalen Zentrums der Mechanischen Wissenschaften (CISM) in Udine, in dem er bis zu seinem Tode leitend tätig war, hatte er maßgeblichen Anteil.

Von seiner überaus umfangreichen wissenschaftlich-publizistischen Tätigkeit zeugen nicht nur seine mehr als 250 Veröffentlichungen in Form von Büchern und Zeitschriftenaufsätzen, sondern auch die durch ihn in den Jahren 1949 und 1953 erfolgte Gründung zweier wichtiger Organe der PAN, des „Archives of Applied Mechanics" und der Forschungshefte „Rozprawy Inżynierskie", die Herausgabe wichtiger IUTAM- und IASS-Kongreßberichte durch ihn, die beratende Tätigkeit als Mitherausgeber anderer Zeitschriften wie des „Journal de Mécanique" (Paris), der „Acta Mechanica" (Wien), des „International Journal of Non-Linear Mechanics" (London) und der vorliegenden Zeitschrift für Angewandte Mathematik und Mechanik (ZAMM) und nicht zuletzt seine zahlreichen Vortragsreisen, die ihn in viele Länder der Welt führten. Er hat an vielen internationalen Kongressen teilgenommen und sie durch Hauptvorträge bereichert und war Organisator und Präsident des IUTAM-Symposiums über „Non-Homogeneity in Elasticity and Plasticity" (1958) und des IASS-Symposiums über „Non-Classical Shell Problems" (1963). Vor 27 Jahren gab er die Anregung zur Veranstaltung der Polnischen Konferenzen über Festkörpermechanik, die sich inzwischen zu einem regelmäßigen wissenschaftlichen Ereignis von internationalem Rang entwickelt haben.

Die Persönlichkeit WACŁAW OLSZAK hatte dank der ihm eigenen Liebenswürdigkeit, seinem Charme und seinem Taktgefühl seiner Fähigkeit, in allen Fragen das Prinzipielle zu erkennen, seiner Gewandtheit, Streitigkeiten zu schlichten, und nicht zuletzt dank seinem unerschöpflichen Humor eine hohe Ausstrahlungskraft. Er betrieb nicht nur selbst eine intensive Forschung, sondern verstand es auch, andere zu inspirieren und für von ihm erkannte Probleme zu begeistern. Er zeichnete sich durch eine hohe gesellschaftliche Kultur aus. Er beherrschte viele Sprachen und vermochte es, seine Gedanken in geschliffenen Formulierungen und in einem gepflegten Stil auszudrücken.

Durch den Tod WACŁAW OLSZAKS hat die polnische und hat die internationale Wissenschaft einen schweren Verlust erlitten. Für die Entwicklung der Mechanik in Polen bedeutet er den Abschluß einer wichtigen Epoche. Zahlreiche Schüler und Kollegen in Polen trauern um ihn, mit ihnen viele Freunde in der ganzen Welt, deren gemeinsames Streben auf die Weiterentwicklung der Wissenschaft, auf die friedliche Zusammenarbeit der Wissenschaftler der Welt und auf die Heranbildung neuer Generationen von Wissenschaftlern hinzielt.

W. FISZDON, A. SAWCZUK

FSC
www.fsc.org
MIX
Papier | Fördert
gute Waldnutzung
FSC® C083411

Zeitfracht Medien GmbH
Ferdinand-Jühlke-Straße 7
99095 Erfurt, Deutschland
produktsicherheit@kolibri360.de